ADVANCED LEVEL CHEMISTRY FOR LIFE

Valerie Moseley

UNIT 1

ADVANCED LEVEL CHEMISTRY FOR LIFE

Title: Advanced Level Chemistry For Life

First Edition: [June 2017]
Second Edition: [January 2018]
Third Edition [June 2019]
Published by VHM Publishing
Printed in Barbados
ISBN: [978-976-96047-0-4]

About the Author

Mrs. Valerie Moseley is a graduate of the University of the West Indies Mona. She graduated with B.Sc. honours in Zoology, Botany and Chemistry in 1976. She holds a Diploma in Education from the University of the West Indies Cave Hill campus.

Mrs. Moseley taught chemistry at the Saint Michael's Girls School in Barbados and at Jamaica College between 1976 and 1977. She then took up an appointment at Queen's College, Barbados in 1978 and retired in 2007. At Queen's College she taught Chemistry and Biology at both ordinary and advanced levels preparing students for the General Certificate of Education (GCE), the Caribbean Secondary Education Certificate (CSEC) and the Caribbean Advanced Proficiency Examination (CAPE). She headed the Physical Sciences department (Physics and Chemistry) and later, the Chemistry department at Queen's College between 1993 and 2007.

Mrs. Moseley has been involved with the Caribbean Examination Council (CXC) in different capacities: as a member of the CAPE syllabus Committee, as a CSEC Examiner moderating the School Based Assessment, and as a CAPE Examiner, in Chemistry. She also conducted CSEC and CAPE School Based Assessment (SBA) Chemistry Practical workshops in some Caribbean islands on behalf of the Caribbean Examination Council (CXC.)

From 2007, Mrs. Moseley has been the Director of Visionary Academy, a Science Institution located in Barbados.

Acknowledgements

I thank the Almighty God who has provided me with the wisdom and the understanding to be able to produce this book. Thanks to Apostle David Coulthrust, who activated the idea of writing this book that had remained dormant in my mind for several years.

Thanks to my many students from whom I gained pleasure as a teacher and joy in developing them as young scientists, over the years. It is their many questions and different learning styles that have challenged my teaching methodology and birthed my desire to present chemistry as a subject for life. I extend special thanks to students Ashley Gibson, Kia Barrow and Danielle Grazette for their assistance in getting this book off the ground.

Thanks to Ms. Nichole Murray of Merit International Co. Ltd who met for many hours, typed the book and produced the graphics.

Thanks to Ms. Patricia Murray, Chemistry teacher at Queen's College for proof reading the first draft of this book.

Thanks to Mrs. Annette Maynard-Alleyne. She went through the final draft of this book very thoroughly and edited. She herself is a co-author of an Advanced Chemistry Study Guide. She is a Chemistry Tutor at The Barbados Community College.

Thanks also to Camella Riley-Pilgrim for the formatting of the book and uploading it to Amazon.

Last but not least, thanks to my family who has been my constant source of inspiration.

Contents

Preface

Chemistry is an applied science. Chemistry is all around us, in everything we do and experience from sports to cooking in the kitchen; from gardening and agriculture to manufacturing and the environment. Chemistry is in medicine, life processes and technological processes. Chemistry is in hairdressing and the beauty industry and also in house cleaning.

Students of Chemistry should be able to apply their knowledge and understanding of the principles of Chemistry to unfamiliar situations. They should be able to create devices and procedures that will help to solve problems in daily life. Chemistry should help students to design plans and execute research in order to benefit mankind.

This book is a student-centred and student friendly, teaching and learning tool that is designed to make seemingly difficult subject material easy to grasp and understand. The language is simple, concise and precise. There are many worked examples of problems and there are questions at the end of most chapters. Students must supplement this book with questions from past examinations, where possible.

Advanced Chemistry for Life is personalised, intended to be the student's actual notebook, with provision for questions to be answered in the book. This is the first of two units and it is patterned off of the CAPE syllabus and is appropriate for students sitting Advanced Level Chemistry in Grade 11 and lower sixth forms in the Caribbean, Africa, Britain and the USA.

Advanced Level Chemistry for Life covers the fundamental principles of Chemistry, Kinetics and Equilibria and the Inorganic Chemistry of selected Groups and Periods in the Periodic Table.

Unit II covers Organic Chemistry, chemistry of Analytical Processes, as well as Environmental and Industrial Chemistry.

<u>Chapter 1</u>

DALTON'S ATOMIC THEORY AND THE STRUCTURE OF THE ATOM

Dalton's Atomic Theory - 1803

Matter is made up of particles. These particles are atoms, ions or molecules. Matter can exist in the solid, liquid or gaseous state and can be an element or a compound.

Postulates/Assumptions of Dalton's Atomic Theory

1. All elements are made up of very small particles called atoms.
2. Atoms can neither be created nor destroyed.
   That is, when matter decomposes, the atoms are recovered unchanged.
3. Atoms combine chemically in small whole number ratios to form compounds.
   Eg. CO_2 is a 1:2 ratio of carbon to oxygen atoms.
4. Atoms are indivisible.
   This means that they cannot be divided into any simpler particles.
5. Atoms of the same element are identical.
6. Atoms of different elements have different masses and different chemical properties.

What are Theories?

Theories are put forward on the basis of hypotheses that are proven consistently to be correct. A hypothesis is a 'hunch' or 'idea' based on some observations or experience. If the hypothesis is tested and proven to be correct, it becomes a theory.

On what basis are theories accepted?

1. There must first be evidence that proves the theory. For example, one cannot say that the MMR vaccine causes autism disorder unless there is ample, provable evidence.
2. The experiments that are conducted to prove the theory must be repeatable and reproducible. This means that identical results must be obtainable by different scientists who carry out the same procedure under similar as well as different conditions.
3. The data or results collected must be accurate and reliable and there must be no manipulating or manufacturing of results.
4. To this end, there must be agreement or consensus on the data within the scientific community. Therefore, Scientist 'A' should not come to one conclusion and Scientist 'B' another about the same issue. They must all agree for the theory to be accepted.

How do Theories Impact on Society?

Theories cause **changes of lifestyle**, for example, in health practices.

The findings of the scientific community on the impact of smoking on health save lives, lower health costs to government, decrease cigarette production and therefore loss of jobs.

In another example, research on alcohol consumption has led to an improvement in health.

Discuss some social and economic changes, in your region that have been impacted by scientific research and emerging theories.

Assumptions of Dalton's Atomic Theory that were proven by other Scientists to be incorrect:

1. Atoms are indivisible.

Atoms can be divided into protons, neutrons and electrons.

Work by Eugen Goldstein in 1880 and Ernest Rutherford in 1911 led to the discovery of protons. In 1897 J. J. Thomson discovered electrons and in 1932 James Chadwick discovered neutrons.

2. Atoms of the same element are identical.

Most elements have different atoms with different numbers of neutrons. These are called isotopes.

An Atom is the smallest part of an element or a compound that can take part in a chemical reaction.

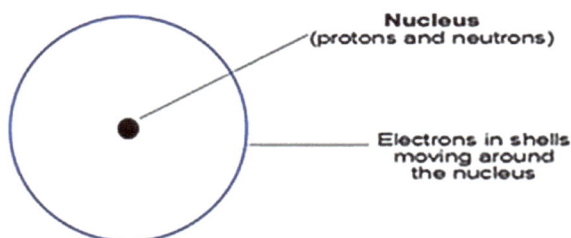

Fig.1a *Structure of the Atom*

	Charge	Mass (kg)	Relative mass	Effect of electric and magnetic fields
Protons	+	1.67×10^{-27}	1	
Neutrons	0	1.67×10^{-27}	1	
Electrons	–	9.1×10^{-31}	1/1840	

Table 1.1 *Comparing properties of protons, neutrons and electrons*

Atomic number –This is the number of protons in the nucleus of an atom.

Note that the number of protons equals the number of electrons and because they are oppositely charged, it confers electrical neutrality on the atom.

Mass number – This is the sum of the nucleons (protons + neutrons)

Isotopes – are different atoms of the same element with different numbers of neutrons. Since all the atoms of that element would have the same number of protons but different numbers of neutrons, then isotopes have different mass numbers. Most elements have isotopes.

For example, the element hydrogen has three isotopes: $^{1}_{1}H$ – hydrogen has 0 neutrons.

$^{2}_{1}H$ – deuterium has 1 neutron and $^{3}_{1}H$ – tritium has 2 neutrons.

They all have 1 proton.

Carbon also has three isotopes: $^{12}_{6}C$ has 6 protons (p) and 6 neutrons (n).

$^{13}_{6}C$, has 6 protons and 7 neutrons while $^{14}_{6}C$ has 6 protons and 8 neutrons.

Relative masses: These are not absolute masses but are compared to a standard. The C-12 isotope is used as the standard. They have NO UNITS.

Relative Atomic Mass – A_r: The ratio of the average mass of the atoms of an element to 1/12th the mass of the Carbon-12 isotope.

Relative Isotopic mass – I_r: The ratio of the mass of an isotope of an element to 1/12th the mass of the C-12 isotope.

Calculating the average Relative Atomic Mass

Since elements have atoms with different numbers of neutrons (isotopes), then the relative atomic mass of the element must be an average value. To calculate the average, information on the following must be known:

- The number of isotopes of the element
- The isotopic masses
- The relative abundance of each isotope /proportion or percentage.

A mass spectrometer is an instrument used to obtain this information.

Fig.1b and Fig.1c below are examples of mass spectrograms.

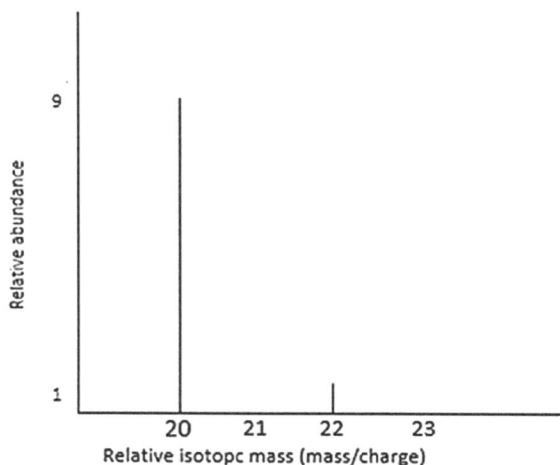

Fig.1b *Mass Spectrum of an element 'A'*

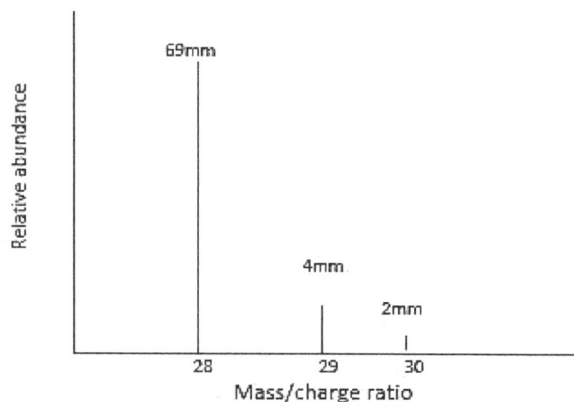

Fig.1c *Mass Spectrum of an element 'D'*

Exercise 1.1

Fig. 1b shows the mass spectrum of an element A.

a. i) How many isotopes does element A have? (1 mark)

_____

ii) State the isotopic mass of each isotope. (2 marks)

_____

iii) Comment on the relative amount of each isotope present in element A. (2 marks)

_____

_____

Worked example 1.1

Calculate the average relative atomic mass of element A.

Solution

$A_{r(av)} = \sum$ *isotopic mass x relative abundance*

= (20 x 9 / 9+1) + (22 x 1/10)

= 20 x 0.9 + 22 x 0.1

= 18.0 + 2.2 = **20.20** (note: two decimal places and no unit)

Worked example 1.2

The element chorine has two isotopes. They are chlorine 37 (^{37}Cl) and chlorine 35 (^{35}Cl). The average relative atomic mass of chlorine is 35.5. Determine the relative abundance of the two isotopes.

Solution: *substitute in* $A_{r(av)} = \sum$ *isotopic mass x relative abundance*

Let **y** = rel. abundance of Cl-35

Then, **1-y** = rel. abundance of Cl-37

4

35.5= (35x y) + (37 x **1-y**)

35.5= 35y + 37 – 37y

35.5 – 37 = 35y – 37y

2X = 1.5

Therefore, y = ¹·⁵/₂ = .75   Chlorine-35= .75   Chlorine-37= 1 - .75 = .25

Worked example 1.3

a.   Use Fig. 1c to calculate the $A_{r(av)}$ of the unknown element 'D'.

Solution: To obtain relative abundance, measure the height of each peak and substitute in

$$A_{r(av) =} \sum isotopic\ mass\ x\ relative\ abundance$$

$$= (^{69}/_{75} \text{ x } 28) + (^{4}/_{75} \text{ x } 29) + (^{2}/_{75} \text{ x } 30) = \textbf{28.11}$$

b.   Using the periodic table, determine the identity of the element 'D'.

_____

Exercise 1.2
An element 'R' gives the following mass spectrum data:

C-46 has a relative abundance of 8.0%.
C-47 has a relative abundance of 7.3%.
C-48 has a relative abundance of 73.4%.
C-49 has a relative abundance of 5.9%.
C-50 has a relative abundance of 5.4%.

a. How many isotopes does 'R' have? (1 mark)

_____

b. Comment on the relative abundance of these isotopes. (2 marks)

_____

_____

c. Calculate the average relative atomic mass of element 'R'. (2 marks)

d. What is the identity of element 'R'? _____

Chapter 2
THE NUCLEAR STRUCTURE OF THE ATOM AND RADIOACTIVITY

Radioactivity is the emission of radiation from the nucleus of an atom. There are two types of radioactivity: Natural and Artificial.

Natural radioactivity is the spontaneous emission of radiation from the nucleus of an unstable atom. Isotopes that emit radiation spontaneously are described as unstable and radioactive. An example is C-14.

Artificial radioactivity is the emission of radiation on bombarding an atom's nucleus with small highly energetic particles for example, protons and neutrons.

Scientists associated with the discovery of radiation are Henri Becquerel (1896) and Pierre & Marie Curie (1898).

Radioactive isotopes emit 3 types of radiation. These are Alpha α, Beta β and Gamma γ.

Alpha α particles are:

- Fast moving helium nuclei ($^{4}_{2}He_{2+}$)
- Positively charged and deflected by electric and magnetic fields.
- Symbol $^{4}_{2}He$

Beta β particles are:

- Electron like particles
- Move rapidly up to the speed of light
- Have the symbol $^{0}_{-1}e$
- Formed when a neutron disintegrates to produce a proton
$$^{1}_{0}n \rightarrow {^{0}_{-1}e} + {^{1}_{1}p}$$
- Beta emissions increase the number of protons

Gamma γ radiation

- Highly energetic, high frequency electromagnetic waves
- Wavelength is approximately one million times shorter than that of visible light
- Gamma ray emission usually accompanies the loss of beta particles

	type		
	Alpha α	**Beta β**	**Gamma γ**
Nature/symbol	Helium nuclei,/ ^4_2He	Electron–like/ $^0_{-1}e$	Electromagnetic radiation of a very high frequency
Mass/	4	$\approx \dfrac{1}{1840}$, which is negligible	no mass
Charge	+2	-1	No charge
Range *	Several cm	Several m	Several km
Relative penetration power	1 (penetrates paper)	100 (penetrates Aluminium foil)	10000 (penetrates Pb)
Absorbed by	Paper, air	Aluminium sheet	Several cm of lead, several metres of concrete
Deflected by an electric or magnetic field	Yes	Yes. deflected greatly	No

Table 2.1 *Comparing the properties of Alpha α, Beta β and Gamma γ radiation*

* *Refers*   *to distance travelled through air.*

Writing and Balancing Equations for Nuclear Reactions

Equations must be balanced in terms of charge, atomic number and mass number.

Equations for alpha decay / emission

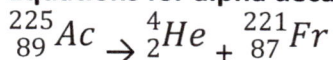

$$^{225}_{89}Ac \rightarrow\ ^4_2He +\ ^{221}_{87}Fr$$

Beta emission

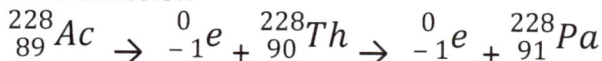

$$^{228}_{89}Ac \rightarrow\ ^0_{-1}e +\ ^{228}_{90}Th \rightarrow\ ^0_{-1}e +\ ^{228}_{91}Pa$$

Note. *Sometimes a series of disintegrations must occur before a stable nucleus is formed.*

Exercise 2.1
Complete the following nuclear equations:

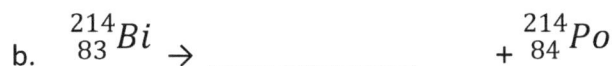

a. $^{228}_{90}Th \rightarrow {}^{4}_{2}He +$ _____

b. $^{214}_{83}Bi \rightarrow$ _____ $+ {}^{214}_{84}Po$

Stable and unstable nuclei

The rate of radioactive decay is measured by the half-life ($T_{1/2}$) of the reaction.
Half-life is defined as the time taken for the concentration of the radio isotope to decrease to half of its original value.

The longer the half-life, the slower the rate of decay and the more stable the isotope is.

Radioisotope	Half life
$^{238}_{92}U$	4.5 X 10^9 years
$^{14}_{6}C$	5.7 x 10^3 years
$^{90}_{38}Sr$	28 years
$^{214}_{83}Bi$	19.7 minutes
$^{214}_{84}Po$	1.5 x 10^{-4} seconds

Table 2. *Half-lives of some radioisotopes*

Worked example 2.2
a. i) What mass of a 10g sample of Strontium-90 remains after 112 years?

Solution

time /yrs.	mass remaining/g
0	10
28	5
56	2.5
84	1.25
112	**0.625g**

Or

112/28 = 4 half lives
1/2 x1/2 x1/2x1/2x10 = 1/16 x10 = **0.625g**

Exercise 2.2

a. How long will it take for 1 gram of U-238 to decrease to 3.125×10^{-2} g? (2 marks)

_____

_____

_____

b. Explain how the C-14 isotope may be used to determine the age of fossils. (3 marks)

_____

_____

_____

What influences the stability of radioisotopes?

The stability of radioisotopes is influenced by **the neutron to proton ratio (n:p) in the nucleus of the atom**.

For small isotopes a 1:1 neutron to proton ratio is most stable.

As the isotopic mass increases, the number of neutrons increase over the number of protons.

The n:p ratio of stable nuclei lies within a restricted band called **the band of stability**. See Figure 2a.

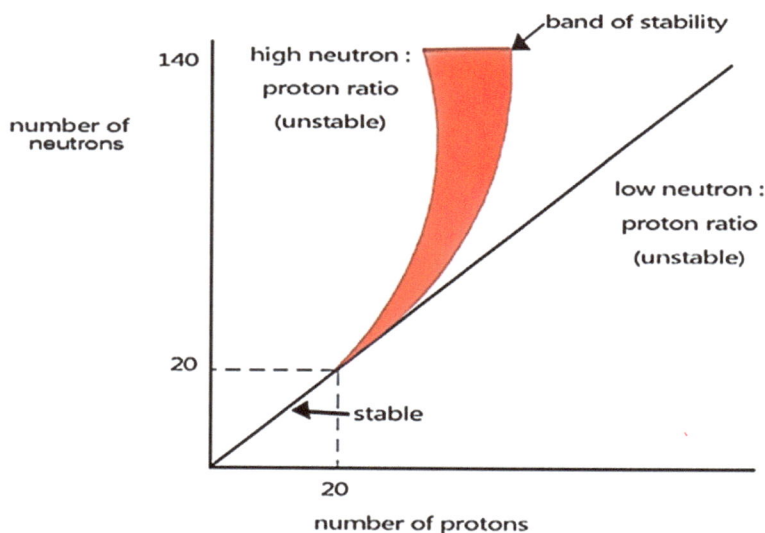

Fig. 2a _Graph showing neutron (n): proton (p) ratio of stable_

The n: p ratio to the left of the band of stability is high.

In order for the isotopes that have a high n:p ratio to produce stable isotopes that fall in the band of stability, the number of neutrons must decrease and the number of protons increase.

This is achieved by the **Emission of *beta* particles** as follows:

For example $\quad {}^{14}_{6}C \quad \rightarrow \quad {}^{14}_{7}N \quad + \quad {}^{0}_{-1}e$

$\qquad\qquad$ 8n : 6p $\qquad$ 7n:7p

$\qquad\qquad$ 1.33: 1 $\qquad$ 1 : 1 ratios

Heavy isotopes must undergo a series of beta emissions before achieving a n:p ratio that lies in the band of stability.

The n:p ratio to the right of the band of stability is low

These undergo **electron capture** or **α decay** in order to increase the number of neutrons and decrease the number of protons. An electron is captured from the innermost shell of the nucleus and it converts a proton to a neutron.

$${}^{1}_{1}p + {}^{0}_{-1}e \rightarrow {}^{1}_{0}n \text{ (electron capture)}$$

This process of electron capture continues until an isotope is produced whose n:p ratio falls on the straight line or in the band of stability

Electron Capture from the innermost shell by the nucleus

For example, $\qquad {}^{119}_{52}Te \quad + {}^{0}_{-1}e \rightarrow {}^{119}_{51}Sb \quad + \quad {}^{0}_{-1}e \rightarrow {}^{119}_{50}Sn$

n:p ratio $\qquad$ 1.29:1 $\qquad\qquad$ 1.33:1 $\qquad\qquad$ 1.38:1

Emission of alpha particles

For example, $\qquad {}^{207}_{81}Tl \quad \rightarrow \quad {}^{4}_{2}He + {}^{203}_{79}Au \quad \rightarrow \quad {}^{4}_{2}He + {}^{199}_{77}Ir$

n:p ratio $\quad$ 1.56:1 $\qquad\qquad$ 1.57:1 $\qquad\qquad$ 1.58:1

Uses of Radioisotopes

1. **Carbon dating of fossils**

   Knowledge of the concentration of C-14 in living tissue and in the fossil, as well as the half-life of C-14 (5700 years), allows for the age of the fossil to be calculated.

2. **Tracers**

   By 'labelling' a molecule with a radioactive isotope the movement of that atom around the body or through a plant can be followed or traced. An example is, in determining the source of the O atom in $C_6H_{12}O_6$ produced during photosynthesis, that is, whether from water (H_2O) or carbon dioxide (CO_2). The radioactive ^{18}O atom in CO_2 or H_2O is followed and traced.

3. **Medical Therapy**

   Gamma rays produced from Cobalt-60 will destroy cancer cells in inaccessible tumours.

4. **As an Energy Source**

- Nuclear energy is widely used in France, Russia, Japan and other parts of the world to generate electricity.
- Plutonium-238 is also used as a source of energy is some heart pacemakers.
- Nuclear Weapons produce so much energy that they can destroy structures several kilometres away.

Disposal of nuclear waste

In countries like Russia, France and the USA, there are nuclear plants for generation of electricity. The nuclear waste from these plants continues to emit harmful radiation and many of the isotopes have very long half-lives and will take thousands of years before emission ceases. Proper disposal of nuclear waste must ensure the following:

- Animals and plants are not exposed to the radiation
- Radioactive waste does not enter the food chain or leak into the water supply
- Geological events, for example earthquakes, do not affect the storage.

The most widely used method of disposal is storage in steel-lined concrete pools 5 meters thick filled with water. Highly radioactive waste is stored in stainless steel tanks embedded in thick concrete jackets and these may then be buried deep underground. Methods being explored for the future include storage under the sea bed and in space.

Exercise 2.3
a. Write balanced nuclear equations for each of the following changes:
   i)   Beta emission from magnesium-28 (1 mark)

_____

   ii) Electron capture by argon-37 (1 mark)

_____

   iii) Alpha emission from plutonium-242 (1 mark)

_____

b. i) Explain the significance of the band of stability? (2 marks)

_____

   ii) Consider potassium-38. Is this a stable or an unstable nucleus? Explain your answer.
       (3 marks)

_____

_____

   iii) If unstable, what radioactive processes must occur to produce a stable nucleus? Write
        relevant nuclear equations to show the reactions. (2 marks)

_____

c. Discuss the advantages and disadvantages of using nuclear energy as a source of electricity compared to the use of fossil fuels. (3 marks)

Chapter 3

THE ELECTRONIC STRUCTURE OF THE ATOM

Radioactivity and nuclear reactions are a property of the atomic nucleus. Chemical reactions are the property of the electrons that are moving around the nucleus. Chemical behaviour depends on the number and the arrangement of the electrons. For example, C-14 is radioactive but C-12 is not, but both have identical chemical properties because they have the same number of electrons.

Experimental findings explaining how electrons are arranged

Observation
It was observed that atoms and molecules emit light when supplied with energy from a source, for example, heat or electricity. The light emitted, when viewed through a spectroscope, has a number of coloured lines called an emission spectrum.

Explanation
In 1913 Bohr explained the emission spectra by suggesting the following:
- Electrons move around the nucleus in stable energy levels/ shells and are said to be in their ground state.
- On absorbing energy, the electrons become "excited" and they move from their stable, ground state into higher energy levels where they are unstable and cannot remain.
- As the electron returns to the ground state, energy in the form of light is released.
- Each line in an emission spectrum represents fixed wavelengths of light emitted as the electrons falls to a lower energy level from a higher one.
- The greater the energy difference between the energy levels, the higher the frequency of the corresponding spectral line and the shorter the wavelength.

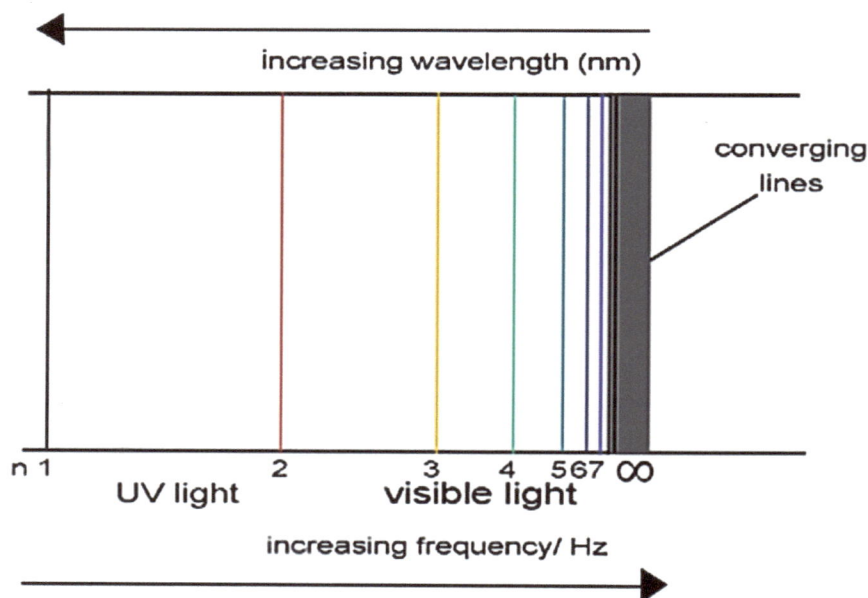

Fig. 3a *Emission spectrum of hydrogen*

Conclusion

1. An atom has an infinite number of energy levels or shells, which are represented by the lines in the spectrum above.
2. Each energy level has a specific amount or quantum of energy associated with it, and is assigned a quantum number called the *Principal Quantum Number* (n).
3. The higher the energy level, the greater the value of n and the further its distance from the nucleus.
4. All known atoms in the ground state have electrons in the n=1 to the n=7 energy levels only.

The **Balmer series** is the series of lines from the infinity energy level to the n=2 energy level of the hydrogen atom.

Visible light is emitted by the hydrogen electron down to the n=2 energy level.
Ultraviolet light is emitted from the n=2 to the n=1 energy level.

Lyman studied the fall of the hydrogen electron back to the ground state n=1, hence, the series of lines down to the n=1 energy level is called the **Lyman series**. The **Paschen series** and the **Brackett series** show transitions down to the n =3 and n=4 energy levels respectively, as shown in Figure 3b.

The energy difference (ΔE) can be calculated from the frequency of the corresponding spectral line.

$\Delta E = E_{ex} - E_g = h\nu$ per atom where 'ex' means excited and 'g' means ground state.

h, Planck's constant = 6.626×10^{-37} kJ Hz^{-1} per atom and ν = frequency/Hz
OR
$\Delta E = h\nu L$ per mole of atoms where L = the Avogadro's constant 6.023×10^{23} mol^{-1}

Worked example 3.1
Calculate ΔE between the n=1 and n=2 energy levels for 1 mole of H atoms, given that the line in the Lyman series has a frequency of 2.47×10^{15} Hz.

Solution
$\Delta E = h\nu L$
= 6.626×10^{-37} KJHz^{-1} x 2.47×10^{15} Hz x 6.023×10^{23} mol^{-1} = **985.70 kJ mol^{-1}**

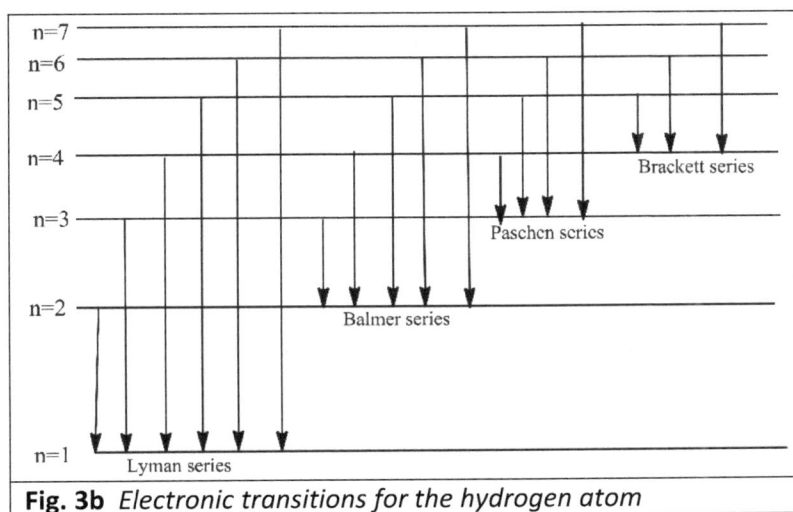

Fig. 3b *Electronic transitions for the hydrogen atom*

The Electronic Configuration of Atoms

Where exactly are electrons found?

Electrons can be thought of, as moving around the nucleus in shells or energy levels which are each assigned a principal quantum number 'n'. The value of n, ranges from 1 to infinity.
Shells have sub-levels called subshells. These subshells are made up of orbitals. The first shell is the n=1 energy level, the second shell is the n=2 energy level and so on. Subshells are designated a secondary quantum number, 'l' and are assigned the letters s, p, d, f, g and h. Each orbital can hold a maximum of two electrons. The 'p' subshell has 3 orbitals (p_x, p_y, p_z). The 'd' subshell has 5 orbitals. The 'f' subshell has 7 orbitals. The 's' subshell is a single orbital.

Shell/n	Subshells	Maximum no. of electrons
1st 1	s □ 2	2
2nd 2	s □ 2 p □□□ 6	8
3rd 3	s □ 2 p □□□ 6 d □□□□□ 10	18
4th 4	s □ 2 p □□□ 6 d □□□□□ 10 f □□□□□□□ 14	32

Table 3.1 *showing sub-division of shells and subshells*

Writing Electronic Configurations and Structures

There are three rules governing the exact ground state configuration of an atom of an element:

1. **Aufbau Principle**

   Electrons in their ground state occupy orbitals in the order of the orbital energy as follows:

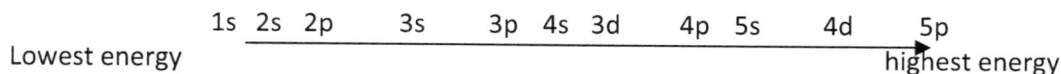

   1s 2s 2p   3s   3p 4s 3d   4p 5s   4d   5p
   Lowest energy ————————————————————————————————————→ highest energy

   Fig. 3c *Showing the order of the energy of the orbitals*

   Orbitals in the same subshell have equal energy and are said to be **degenerate.**

2. **Pauli Exclusion Principle**

   An orbital can be occupied by only 2 electrons at a time and they must have opposite spin.

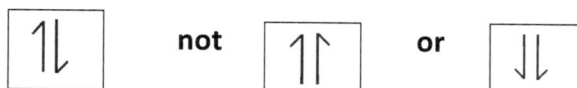

   ⇅ □   **not**   ⇈ □   **or**   ⇊ □

3. Hund's Rule

The orbitals of a subshell must be occupied singly and with parallel spin before they can be occupied in pairs as follows:

not

Worked example 3.3

Write the electronic configurations and structures of the following atoms:

a. $_1$**H**

$1s^1$

b. $_2$**He**

$1s^2$

c. $_{17}$**Cl**

$1s^2 2s^2 2p^6 3s^2 3p^5$

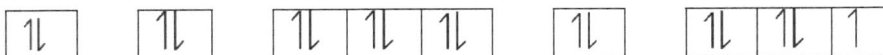

d. $_{22}$**Ti**

$1s^2 2s^2 2p^6 3s^2 3p^6 3d^2 4s^2$

Titanium is a transition metal.

Note *that the 4s orbital is filled before the 3d orbitals. This is characteristic of the transition metals.*

Shapes and Symmetry of Atomic Orbitals

An **orbital** is defined as an area in space where it is most probable that an electron will be found, orbiting around the nucleus. An electron in an s orbital is most likely to be found orbiting the nucleus along the path of a sphere.

The **s orbital** is therefore said to have a **spherical shape** and it is spherically symmetrical.

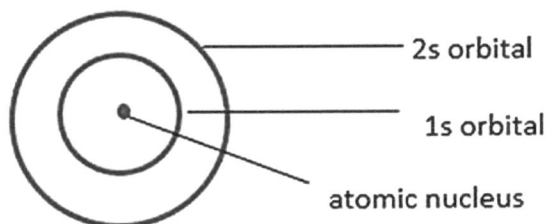

Fig. 3d *Diagram showing the shape and size of the 1s and 2s orbitals*

The electron density of **p orbitals** is equally distributed in two regions on opposite sides of the nucleus. Each p orbital therefore has two lobes and is described as having a **dumb bell shape**. p orbitals are vertical, horizontal and diagonal in their symmetry. These orbitals are arbitrarily given the symbols, p_x, p_y and p_z. The symmetry of the p orbitals is as follows:

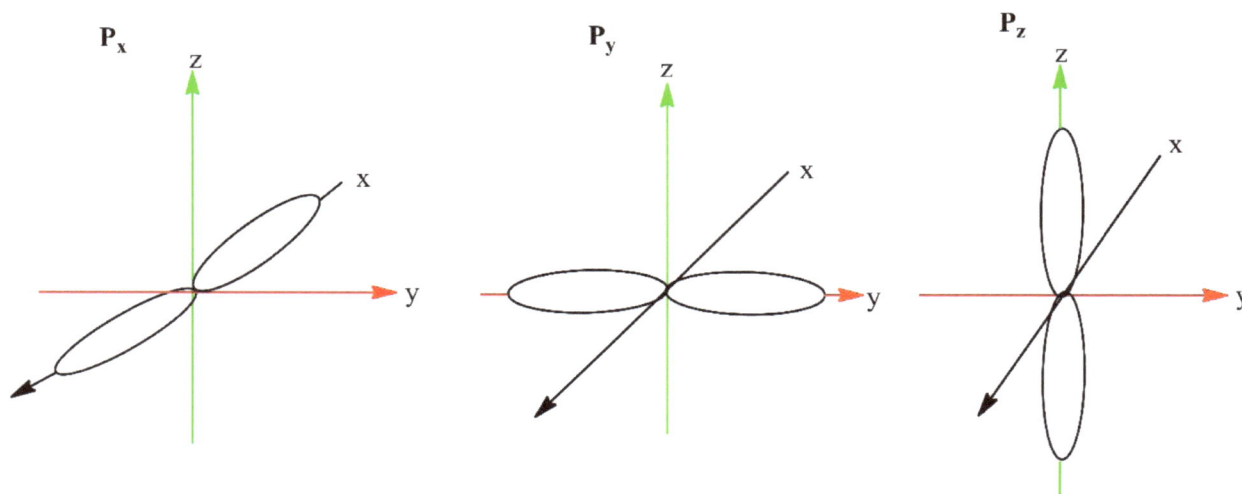

Fig. 3e *Diagram showing the shape and symmetry of p orbitals*

Exercise 3.1

a. Write the electronic configuration and draw the electronic structure for the elements calcium and oxygen. (4 marks)

b. Sketch the shape of the orbitals present in these atoms. (2 marks)

c. The pair of electrons in an orbital is in opposite spin. Explain why the electrons are in opposite spin. (2 marks)

_____

_____

3.2

a. Draw a labelled sketch of the emission spectrum of a hydrogen atom, representing the Lyman series. (3 marks)

b. Explain fully how the lines in this series arise. (3 marks)

_____

_____

_____

_____

c. Calculate the energy in Joules per mole, of a photon of light emitted, having a frequency of 4.0×10^{14} Hz. (2 marks)

Chapter 4

An electron at the infinity energy level where the lines have converged, no longer experiences the attractive forces of the atomic nucleus and escapes from the atom leaving a positively charged ion behind.

$$M_{(g)} - 1e^- \rightarrow M^+_{(g)} ; \Delta H +ve$$

This process is known as **ionisation.** The energy absorbed to bring about ionisation is called **ionisation energy.**

Ionisation energy is defined as the energy required to form 1 mole of gaseous cations, by the loss of one mole of electrons at standard conditions of 1 atm, and 25 °C (298K).

First Ionisation Energy is the energy required to form 1 mole of gaseous uni-positive cations (with a charge of 1+) by the loss of 1 mole of electrons at standard conditions.

$$Al_{(g)} - 1e^- \rightarrow Al^+_{(g)}$$

Second Ionisation Energy is the energy required to form 1 mole of gaseous cations with a charge of 2+ by the loss of 1 mole of electrons at standard conditions.

$$Al^+_{(g)} - 1e^- \rightarrow Al^{2+}_{(g)}$$

Third ionisation energy is the energy required to form 1 mole of gaseous cations with a charge of 3+ by the loss of 1 mole of electrons at standard conditions.

$$Al^{2+}_{(g)} - 1e^- \rightarrow Al^{3+}_{(g)}$$

Periodic Trends in Ionisation Energy

First Ionisation Energy

This depends on the size of the atom, and is expressed in terms of the atomic radius.

The atomic radius is a measure of the distance between the nucleus and the outermost (valence) shell of an atom. It is **defined as** *half the distance between the nuclei of two atoms of the same element, covalently bonded or in a metallic lattice.*

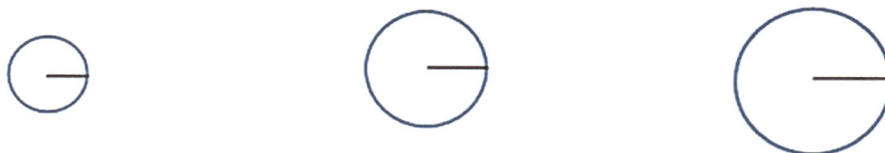

Fig. 4a *Diagram showing increasing atomic radius*

The atomic radius is influenced by 2 main factors:
1. The shielding effect on the valence shell, by inner shells
2. The effective nuclear charge and nuclear charge of the atom.

The Shielding Effect
*   Shells of electrons between the nucleus and the valence shell shield the valence electrons from the nuclear attraction.

- The more strongly the nucleus can attract the valence electrons, the closer the shell is to the nucleus, thus the smaller the atomic radius and the more energy needed to bring about ionisation.

- Increasing number of shells provide a barrier between the nucleus and the valency/ outer shell electrons, resulting in the attractive force of the nucleus getting progressively weaker and the atomic radius increasing and ionisation energy decreasing.

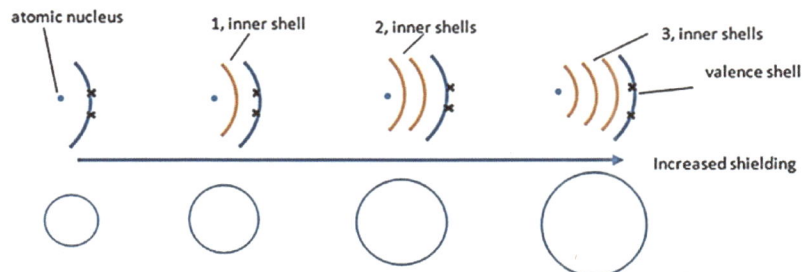

Fig. 4b *Diagram showing the increasing atomic radius as the number of inner shells increases*

Nuclear charge and effective nuclear charge
- The nuclear charge is the total number of protons in the nucleus of an atom.
- The effective nuclear charge relates to the actual number of protons that are exerting attractive forces on the valence electrons.

For example, Sodium ($_{11}Na$) has 11 protons and therefore the nuclear charge is 11. It has 11 electrons distributed in shells as 2.8.1. The 10 protons will attract the 10 inner shell electrons and the single electron in the valence shell is "effectively" being attracted by only one proton. The **effective nuclear** charge of sodium is therefore 1.

- The effective nuclear charge corresponds to the number of electrons in the valence shell, as well as to the group number.
- Both the nuclear charge and the effective nuclear charge increase as you go across the period.
- As nuclear charge/ effective nuclear charge increases, the force of attraction on the valence shell of electrons increases and the *atomic radius decreases*, resulting in an increase in *the ionisation energy*.

Second and successive ionisation energies

The second ionisation energy is higher than the first ionisation energy because the ionic radius is smaller than the corresponding atomic radius. Third, fourth and successive ionisation energies get progressively higher.

The **ionic radii** get smaller because, with fewer electrons, the inter-electron repulsive forces decrease. The remaining electrons are therefore attracted to the nucleus more strongly, decreasing the ionic radius.

Also, as more electrons are lost and the charge on the ion increases, the remaining electrons are attracted more strongly to the nucleus and the ionic radius gets smaller.

The 2nd, 3rd, 4th and successive ionisation energies, therefore, get larger.

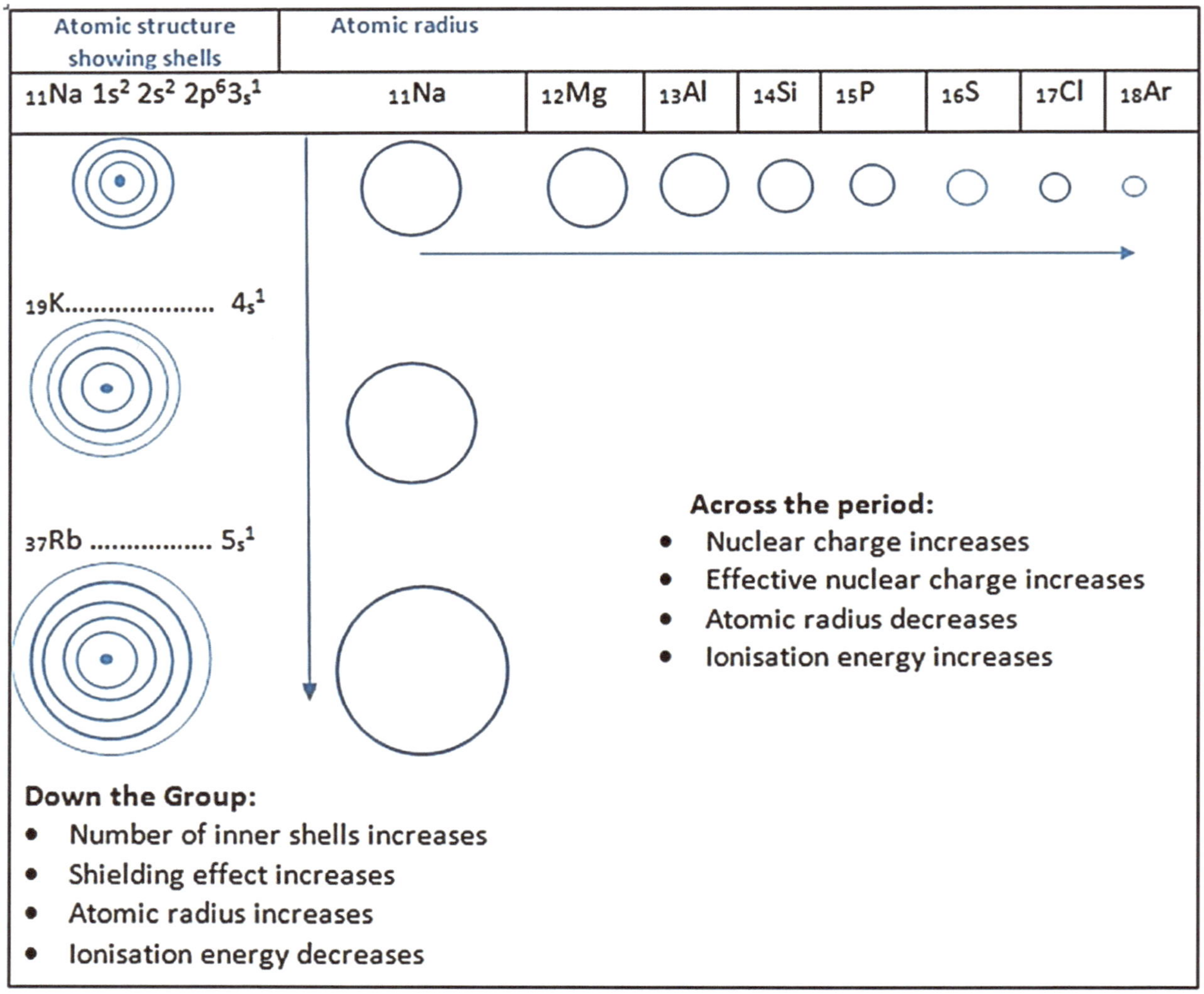

Atomic structure showing shells	Atomic radius							
$_{11}$Na $1s^2$ $2s^2$ $2p^6 3s^1$	$_{11}$Na	$_{12}$Mg	$_{13}$Al	$_{14}$Si	$_{15}$P	$_{16}$S	$_{17}$Cl	$_{18}$Ar

$_{19}$K.................... $4s^1$

$_{37}$Rb .... $5s^1$

Across the period:
- Nuclear charge increases
- Effective nuclear charge increases
- Atomic radius decreases
- Ionisation energy increases

Down the Group:
- Number of inner shells increases
- Shielding effect increases
- Atomic radius increases
- Ionisation energy decreases

Fig. 4c. *Table showing variation in atomic radius across Period 3 and down Group 1*

Other factors that influence the ionisation energies are related to the electronic structure.
- A full orbital is more stable than one which is half-full and therefore requires more energy for ionisation.

Full orbital

- A half-full subshell is more stable than one that is not.

1	1	1

and

1	1	1	1	1

p subshell   **d subshell**

- A full subshell is very stable.
- A full shell is the most stable and requires large amounts of energy for ionisation.

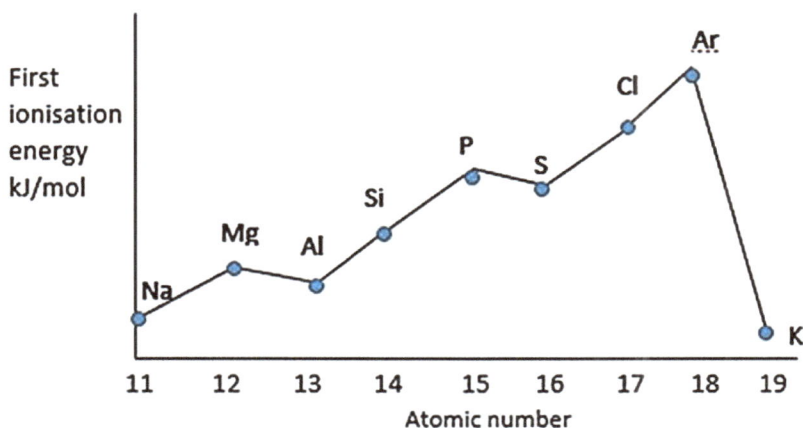

Fig. 4d *Graph of first ionisation energy against atomic number in Period 3 and Group 1*

As you go from sodium to argon, there is a general increase in the first ionization energy as the nuclear charge and effective nuclear charge increase across Period 3. Aluminium and sulphur are anomalous. In the case of aluminium, the first ionization energy is lower than that of magnesium. This is because a single electron is removed from the half-filled $3p_x$ orbital of aluminium, requiring less energy, than for removing an electron from the full 3s orbital of the magnesium atom.

In the case of phosphorus and sulphur the half-filled $3p^3$ subshell of phosphorus is more stable than the $3p^4$ subshell of sulphur. There is less inter-electron repulsion between the 3, phosphorous electrons than the 4 sulphur electrons. The nuclear attraction is therefore stronger in phosphorus than in sulphur. Hence the first ionisation energy of phosphorous is higher than that of sulphur.

The first ionisation energy for potassium is lower than that of sodium. This is because there is an additional shell of electrons in potassium that shields the valence shell from the nuclear attraction, resulting in a larger atomic radius, and a lower first ionisation energy.

Predicting electronic structures from successive Ionisation Energies

By examining the difference between successive ionisation energies, the number of electrons in the valence shell of an atom can be determined, hence the group in which the element is found.

Worked example 4.1

Consider element A with the following successive ionisation energies.

Ionisation energies (kJ mol^{-1})

1st	2nd	3rd	4th	5th
520	1280	4600	5960	7200

Difference   760   3320   1360   1240

Predict the group in the Periodic Table to which Element A belongs and suggest a possible electronic structure.

Solution

As expected the first to fifth ionization energies are increasing. There is a gradual increase from the first to the second ionization energy. However, the large increase in ionization energy to remove the third electron suggests that the third electron is in a stable shell. Hence A has 2 electrons in the outermost shell, has a valence of 2 and it is in group 2.

A possible electronic configuration is as follows: $1s^2 2s^2 2p^6 3s^2$.

Evidence for the existence of shells and subshells from ionization energies

1. The large increase in ionization energy between the second and third ionization energies above suggests that the third electron is in a full and stable shell therefore providing evidence of the existence of shells.
2. Where successive ionization energies gradually increase by small increments, it provides evidence that those electrons are being removed from subshells within a shell.
3. Also as groups are descended the ionisation energy decreases. This provides evidence that increasing numbers of shells are providing a shielding effect.

Exercise 4.1

An element 'R' has the following successive ionisation energies (kJ mol^{-1}).

1st	2nd	3rd	4th	5th	6th
620	1810	3,111	7,650	9,010	11,115

a. i) In which group of the periodic table is element R likely to be found? (1 mark)

_____

ii) Explain your answer. (2 marks)

_____

_____

b. Element 'T' has atomic number 16.

   i) Write the electronic configuration and draw the electronic structure of T in terms of "s p d f" orbitals. (2 marks)

   _____

   ii) In which group of the periodic table is T found? (1 mark) _____

   iii) Using the letters 'R' and 'T' as symbols, write the formula of the compound formed between R and T. (1 mark)

   _____

c. Explain how the data above provide evidence for the existence of shells and subshells in an atom. (3 marks)

   _____

   _____

   _____

Chapter 5

BONDING

Bonding is defined as a force of attraction between oppositely charged particles or entities.

For example

Cations (+) are attracted to Anions (-) in ionic bonding.
Protons (+) are attracted to electrons (-) in covalent bonding and metallic bonding.
Similarly charged particles repel each other.

The types of forces of attraction that are considered in this chapter are:
1. **Covalent bonding**
2. **Dative covalent bonding or co-ordinate bonding**
3. **Intermolecular forces of attraction**
4. **Metallic bonding**
5. **Ionic Bonding**

Covalent Bonding

Covalent bonding is defined as a force of attraction between pairs of shared electrons and their atomic nuclei. It is a very strong bond. It involves the overlap of half-filled orbitals in the valence shell, in order to achieve full orbitals. Molecules are the product of the covalent bond between atoms.
The following examples show the overlap of atomic orbitals during covalent bonding.

1.

$_1H$ + $_1H$ $\longrightarrow$ H_2 molecule

$1s^1$ + $1s^1$

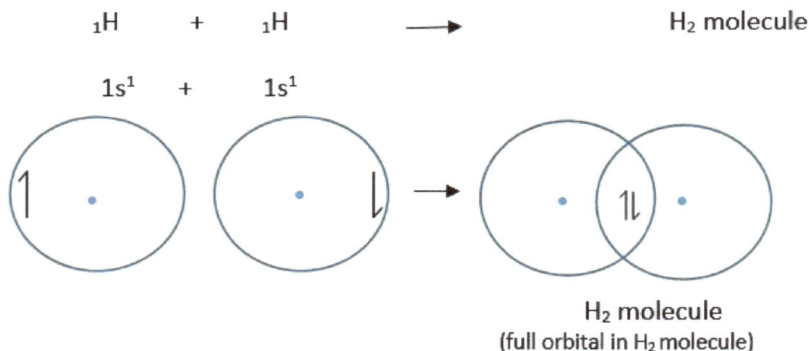

H_2 molecule
(full orbital in H_2 molecule)

2.

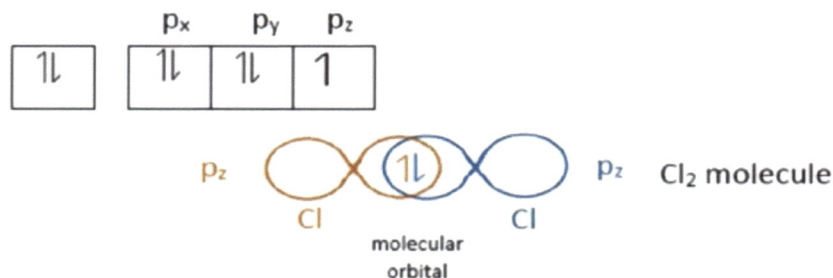

Px Py Pz

Pz Cl Cl Pz Cl_2 molecule
molecular
orbital

3.

$_1H$ + $_{17}Cl$ ⟶ HCl

$1s^1$ + $1s^2 2s^2 2p^6 3s^2 3p^5$ (electronic configuration of H and Cl respectively)

H ⟨↿⇂⟩ Cl   HCl molecule

1s $3p_z$

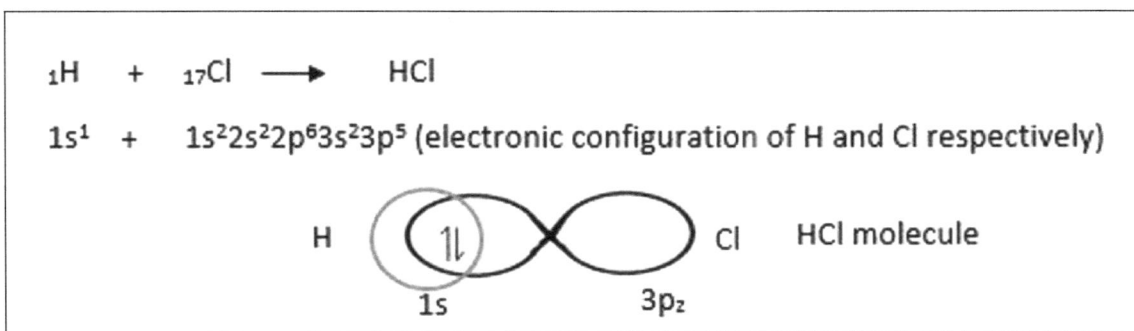

When atomic orbitals overlap molecular orbitals are formed. When the overlap is head on, or end on, the covalent bond is called a **sigma bond (σ)**. Sigma bonds are very strong and very stable covalent bonds with the electron density concentrated between the nuclei of the 2 atoms.

The p orbitals can overlap both head on as well as side on. The head on overlap is the sigma bond and the side on overlap is the **pi bond(π)**. When this occurs, the electron density is concentrated above and below the bond axis. This is a weaker, less stable bond than the sigma bond.

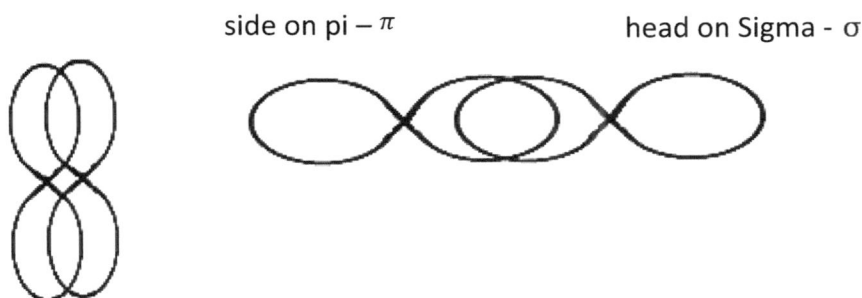

side on pi – π head on Sigma - σ

Usually, covalent bonding occurs between atoms of non-metals, however some metals bond covalently. For these metals, it is too expensive energetically to form stable cations. For example, the elements Be, B, and Al form covalent bonds in some of their compounds. These metals, because of their very small size, require high ionisation energies to form stable 2^+ and 3^+ ions, so they do not generally form ionic compounds. $AlCl_3$ is covalent because the sum of the 3, ionisation energies is too high for Aluminium to form a Al^{3+} ion.

$BeCl_2$ is covalent, whereas $MgCl_2$ is ionic. For Be, the atomic radius is very small, coupled with the need to lose 2 valence electrons to form stable Be^{2+} ions. Similarly, BF_3 is covalent because it also has a small size and requires the sum of the 1st, 2nd and 3rd ionisation energies to form a stable B^{3+} cation.

Hybrid Orbitals

Consider the electronic structure of carbon:

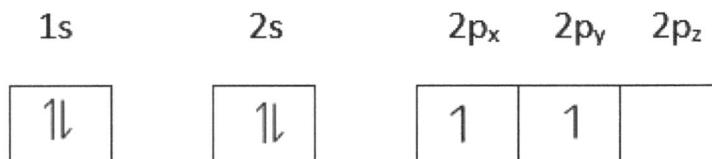

| 1s | 2s | 2p$_x$ | 2p$_y$ | 2p$_z$ |

There are only 2 half-filled orbitals, p$_x$ and p$_y$ with 2 unpaired electrons in the valence shell. Carbon is in group 4 and has a valence of 4, hence 4 unpaired electrons are involved in its bonding. To produce 4 unpaired (single electrons) in the valence shell, a 2s electron on absorbing energy, moves to the next available vacant orbital, which is the p$_z$ orbital.

This is called **promotion**. Now that there are 4 single, unpaired electrons, the orbitals undergo "mixing" to produce 4 equivalent or similar orbitals called **hybrid orbitals**.

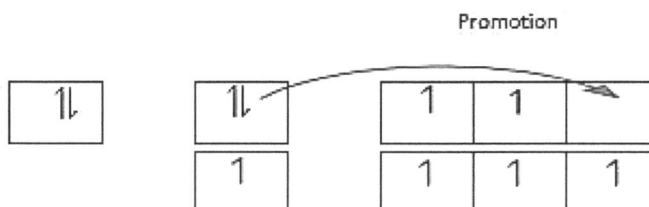

Promotion

If all four orbitals mix, there are 4, **sp³ hybrid orbitals**.
If only three orbitals mix, there are 3, **sp² hybrid orbitals** and **1 p$_z$ orbital**.
If only two orbitals mix, there are 2, **sp hybrid orbitals** and **1 p$_y$** and **1 p$_z$ orbital**.
The shape of the hybrid orbitals is a blend of the s and p orbitals.

Shape of hybrid orbital

Bonding and structure of some organic molecules

1. **Methane - CH₄**

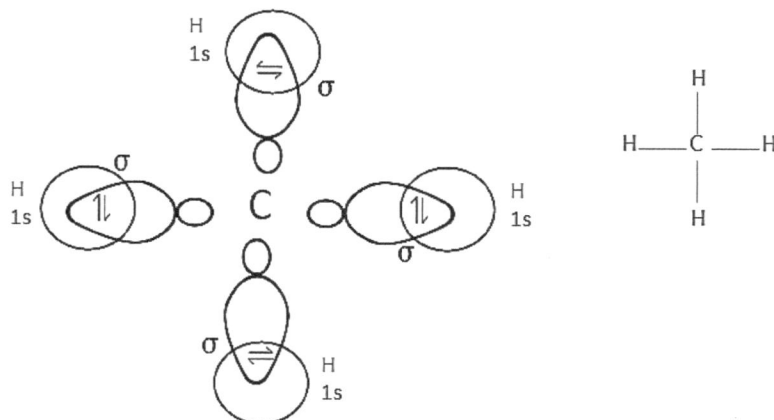

Fig. 5a Diagram showing orbital overlap in the covalent bonding of methane

2.   Ethane - C_2H_6

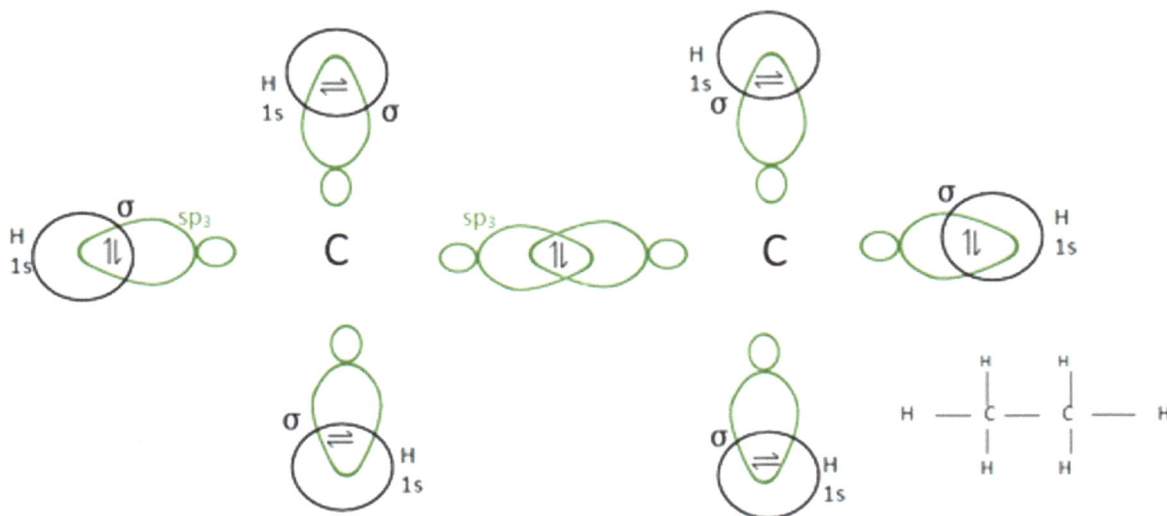

Fig. 5b Diagram showing orbital overlap in the covalent bonding of ethane

3   Ethene - C_2H_4

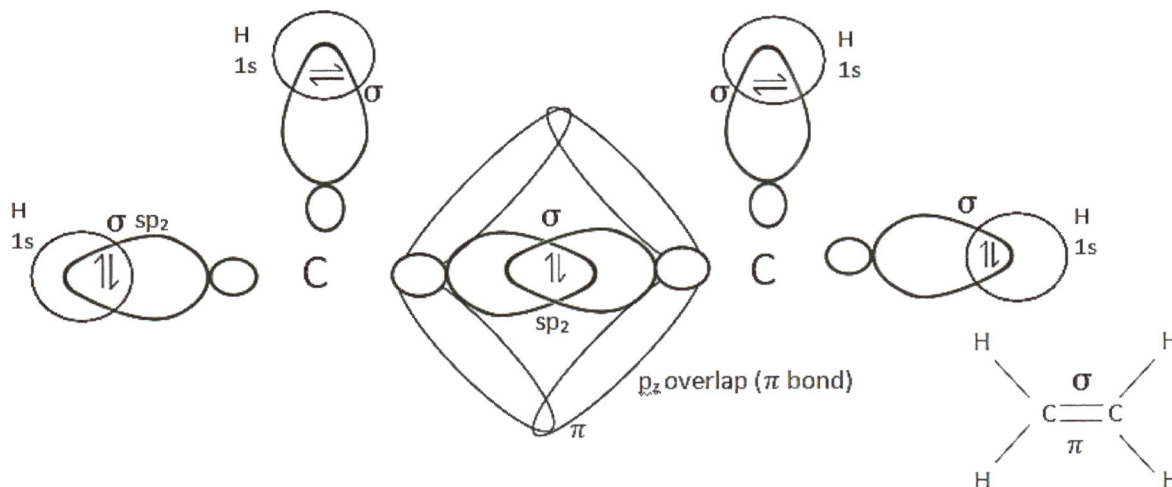

Fig. 5c *Diagram showing orbital overlap in the covalent bonding of ethene*

Bonding and Non-Bonding Electron Pairs

Consider the covalent bonding in the molecule Ammonia (NH_3). The central Nitrogen atom is in group 5 with the following valence shell configuration: $2s^2\, 2p^3$

Notice that the 2s pair of electrons is not involved in bonding with the hydrogen atoms. This pair is called a non-bonding pair or a lone pair of electrons.

Any pair or pairs of electrons in the valence shell of the centrally bonded atom of a molecule, that are not involved in the bonding, are called **non-bonding** or **lone pairs** of electrons.

The strength of the covalent bond increases as the atomic radius of the bonded atoms decreases. For example, as the size of the halogen atom increases down group 7, the covalent bond weakens. Hence the H – F bond is stronger than the H – Cl bond, which is stronger than the H – Br and H – I bonds.

Exercise 5.1
a. Use orbital overlap to show the bonding in (i) N_2 and (ii) $BeCl_2$. (3 marks)

b. Consider the molecule H_2O.
    i) Which atom is the central atom? (1 mark) _____

    ii) How many electrons are in its valence shell? (1 mark) _____

    iii) Use orbital overlap and dot and cross diagrams to show the bonding in the H_2O molecule. (4 marks)

    iv) How many non-bonding pairs of electrons are there in the molecule and where are they found? (2 marks)

5.2
a. Sulphur and oxygen are in the same Group of the periodic table. All of sulphur's valence electrons are bonded in the molecule SF_6.

i) Explain how it is possible for sulphur to produce 6 orbitals with single or unpaired electrons. (3 marks)

_____

_____

ii) Name the orbitals produced. (1 mark) _____

iii) Use dot and cross and orbital overlap to show the bonding in SF$_6$.(4 marks)

Single, Double and Triple Bonds
If one pair of electrons is shared between two atoms in a molecule, the covalent bond is referred to as a **single bond**. If two pairs are shared, it is a **double bond** and if three pairs are shared, it is called a **triple bond**.

Dative Covalent Bonding / Co-ordinate Bonding

This is a type of covalent bonding in which the pair of shared electrons is contributed by 1 atom only. This atom must therefore have at least one lone pair of electrons. The other atom must have a vacant orbital to accommodate the lone pair of electrons.

Co-ordinate bonding often results in the formation of complex molecules and complex ions. For example, transition metal complexes are formed as a result of co-ordinate bonding.

Consider the dative covalent bonding in the ammonium ion:

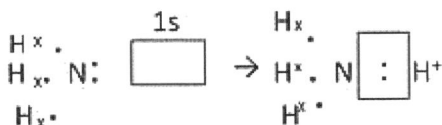

$$NH_3 \ + \ H^+ \ \rightarrow \ NH_4^+$$

Fig. 5d *Dot and cross diagram showing the dative covalent bonding in the ammonium ion*

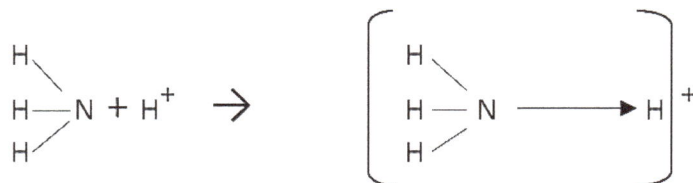

A dative covalent bond forms between AlH_3 and H^- ion as follows:

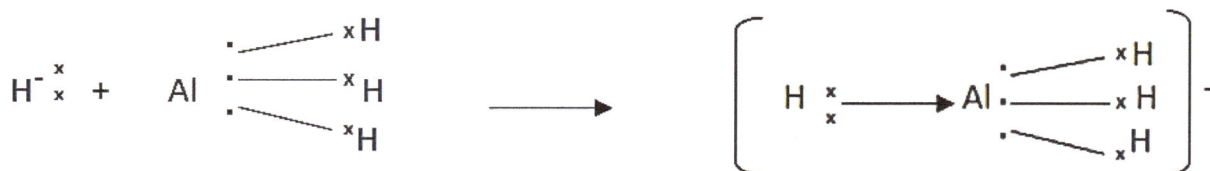

Predicting the Shapes of Covalent Molecules using the Valence Shell Electron Pair Repulsion Theory (VSEPR)

The VSEPR Theory, in its application to predict the three - dimensional shapes of covalent molecules, assumes the following:

- Bonded pairs of electrons repel each other equally.
- Non-bonded pairs of electrons repel bonded pairs more strongly, than the bonded pairs repel each other.
- Non-bonded pairs of electrons repel each other more strongly, than they repel bonded pairs.
- Double bonds repel single bonds slightly more strongly than the single bonds repel each other.

Shapes of molecules with **two bonded pairs** of electrons. For example: **BeCl$_2$**

The central Be atom is in group 2 and therefore has 2 valence electrons. Both of these electrons are involved in bonding in the $BeCl_2$ molecule. It therefore has **2 bonded pairs** of electrons and **0 non-bonded pairs** of electrons. The 2 bonded pairs repel each other equally resulting in $BeCl_2$ having a **linear shape** with **bond angles of 180°**.

$$Cl \text{ —— } Be \text{ —— } Cl$$
$$180°$$

CO_2 is also linear. Double bonds repel each other equally.

$$O = C = O$$
$$180°$$

If there are **3 bonded pairs** of electrons around the central atom and **0 non-bonded pairs**, the bonded pairs repel each other equally to give rise to a **trigonal planar molecule** with **bond angles of 120°**.

For example: **BF₃**.

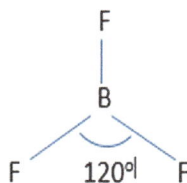

If there are **4 bonded pairs** and **0 non-bonded pairs**, equal repulsion of the bonded pairs results in a **tetrahedral** shaped molecule with **bond angles of 109.5°**.

For example: **CH₄**

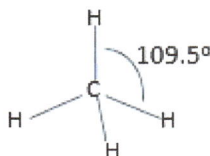

Other examples of tetrahedral molecules are **NH₄⁺** and **CCl₄**.

If the central atom has **5 bonded pairs** and **0 non-bonded pairs**, then equal repulsion of the bonded pairs result in the molecule having a **trigonal bi-pyramid** shape with a mixture of **bond angles** of **90°** and **120°**.

For example: **PH₅**

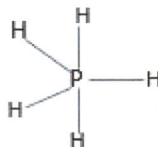

If the central atom has **6 bonded pairs** and **0 non-bonded pairs** repelling each other equally, it gives rise to a molecule that has an **octahedral shape**, with all **90° bond angles**.

Shapes of molecules with non-bonding pairs of electrons

When there are **two bonded pairs** and **1 non-bonded pair** of electrons, around the central atom, the non-bonded pair /lone pair repels the 2 bonded pairs more strongly than the bonded pairs repel each other, resulting in the molecule having a **bent shape**.

For example: **SnCl₂**

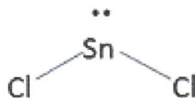

In molecules with **2 bonded** and **2 non-bonded pairs** of electrons around the central atom, the 2 non-bonded pairs repel each other and they repel the bonded pairs more strongly than the bonded pairs repel each other. The molecule therefore has a **'V' shape** and **bond angles of 105°**.

For example: **H_2O**

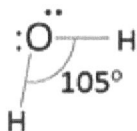

In molecules with **3 bonded pairs and 1 non-bonded pair** of electrons, the 3 bonded pairs repel each other equally, but the non-bonded pair repels them more strongly than they repel each other. This results in the molecule having a **pyramidal shape** with a **bond angle of 107°**:

For example: **NH_3**

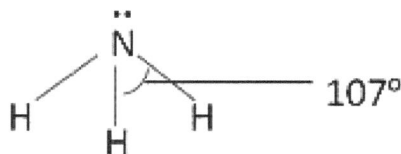

Molecules with **3 bonded pairs** and **2 non-bonded pairs** of electrons have a **T shape** with **bond angles of 180° and 90°**.

For example: **ICl_3**.

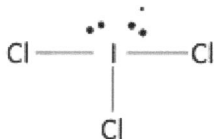

Molecules with **2 bonded pairs** and **3 non-bonded pairs** of electrons have a **linear shape** with **180° bond angle**.

For example: **I_3^-**

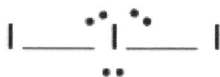

Shapes of some organic molecules:

Ethane - C_2H_6

Equal repulsion of the **4 bonded pairs** of electrons around each central carbon atom, gives a tetrahedral arrangement with respect to each carbon atom.

The molecule is described as a **double tetrahedron** with **bond angles** of **109.5°**.

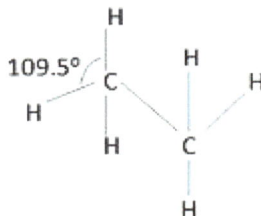

Ethene - C_2H_4

Equal repulsion of the **3 bonded pairs** around each central carbon atom, gives a **trigonal planar** arrangement with respect to each carbon atom. The shape of the ethene molecule is described as **planar** with **bond angles** of **120°**.

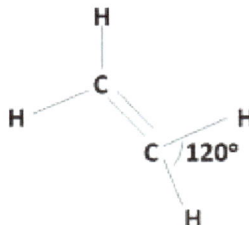

Benzene – C_6H_6

Benzene has a hexagonal arrangement of carbon to carbon bonds in a 6-membered **ring structure**. There is a trigonal arrangement of bonds, with bond angles of 120° with respect to each carbon atom.

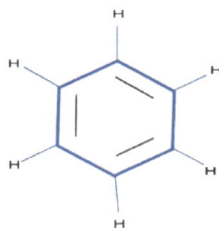

In benzene the pi-bond is delocalised. The benzene molecule is a hybrid of two resonance structures. A **resonance structure** is a molecule with different possible distributions of electron pairs. Benzene has 2 resonance structures. The pi bonds are moving in the Benzene ring. That is to say, they are not localised.

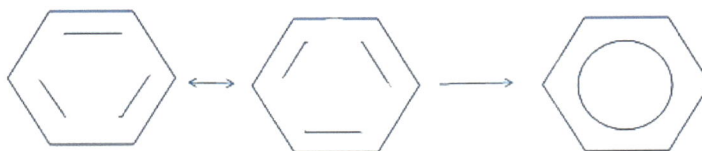

Fig. 5e *Diagrams of the two resonance structures of Benzene*

The ring in the centre of the hexagon represents the delocalised pi electrons.

35

Intermolecular Forces of Attraction

Electronegativity and Polarity

Electronegativity is the tendency for an atom in a covalent bond to attract the electron pair. For example, In the HCl molecule, chlorine is more electronegative than hydrogen and so the pair of electrons between them is unequally shared, such that there is a higher electron density around the chlorine atom. The chlorine atom therefore develops a small negative charge (δ^-) and the hydrogen atom develops a small positive charge (δ^+). The molecule has two poles of charge or **dipoles**. The molecule is said to be **polar**.   **(H $^{\delta+}$- Cl $^{\delta-}$)**

In the Cl_2 molecule (Cl – Cl), the two atoms are identical and are of the same electronegativity and the molecule is said to be **non-polar**.

Intermolecular forces of attraction are very weak forces between molecules. These are very long, weak bonds.

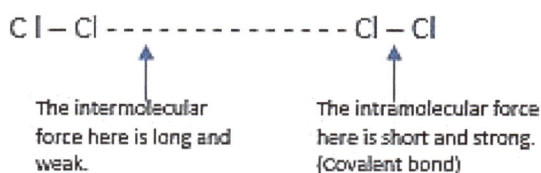

The intermolecular force here is long and weak.

The intramolecular force here is short and strong. (Covalent bond)

Fig. 5f *diagram showing the intra and intermolecular bonding in chlorine*

There are three main types of intermolecular forces of attraction:

1. **Permanent dipole to permanent dipole forces**
2. **Hydrogen bonds**
3. **Temporary dipole to temporary dipole forces (Van der Waal's forces)**

Permanent Dipole – Permanent Dipole Forces

Permanent Dipole forces of attraction occur between polar molecules. Consider HCl, the chlorine (Cl) atom is more electronegative than the hydrogen atom. The Cl atom therefore attracts the bonded electrons toward itself and the electron density will be more concentrated around it than around the hydrogen atom. Two poles of charge are therefore developed.

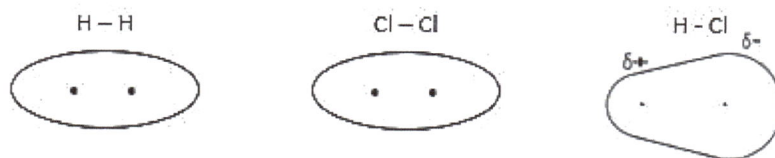

Fig. 5g *Diagram showing non polar H$_2$, and Cl$_2$ and polar HCl*

Under high pressure and low temperature, the HCl molecules are forced together and weak intermolecular forces of attraction develop between the δ^- pole of one molecule and the δ^+ of another molecule. This results in the formation of liquid HCl. This force of attraction is called a **permanent dipole-permanent dipole** interaction.

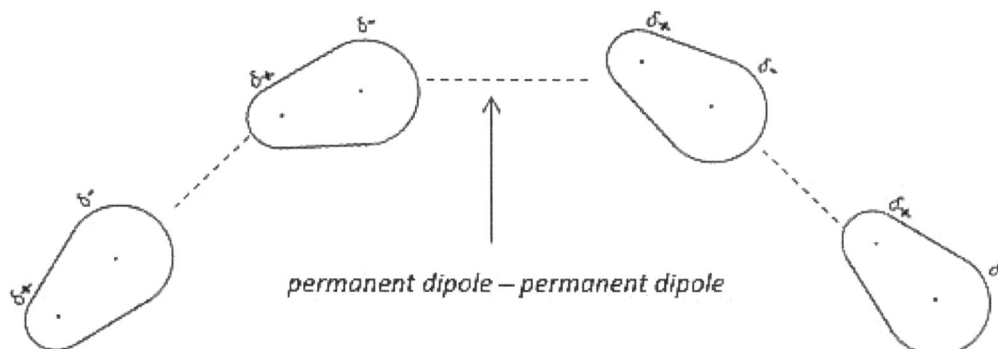

Fig. 5h *Diagram showing the permanent dipole interaction in HCl liquid*

The Hydrogen Bond (H-bond)

The hydrogen bond is a special permanent dipole - permanent dipole interaction. It is defined as a force of attraction between the hydrogen atom in one molecule and a small, highly electronegative atom with lone pairs in another molecule. The three atoms that qualify as being small and highly electronegative are nitrogen, oxygen and fluorine.

The highly electronegative atom in the molecule attracts the bonded pair of electrons so strongly to itself in the covalent bond that it leaves the hydrogen atom a virtual proton. The H $^{\delta+}$ attracts the lone pair in the electronegative atom forming a slightly stronger intermolecular force than the regular permanent dipole - permanent dipole interaction.

Hydrogen bonding occurs in the liquid and solid states of H_2O, HF and NH_3. Each ammonia molecule has one lone pair and forms one hydrogen bond per molecule. Water has two lone pairs and therefore forms two hydrogen bonds per molecule and hydrogen fluoride, which has three lone pairs, forms three hydrogen bonds per molecule.

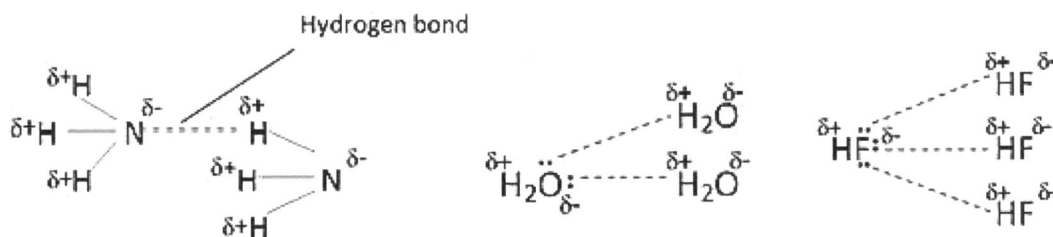

Fig. 5i *Diagram showing hydrogen bonding between molecules of NH_3, H_2O and HF, respectively*

Temporary Dipole- Temporary Dipole forces of attraction (Van der Waal's forces)

Temporary dipole – temporary dipole forces of attraction occur between non-polar molecules and are the weakest intermolecular forces of attraction.

In liquid oxygen, this force bonds oxygen molecules as shown below.

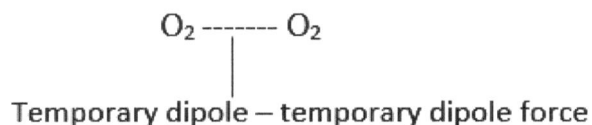

O_2 ------- O_2

Temporary dipole – temporary dipole force

Fig. 5j *A symmetrical oxygen molecule*

Since the two oxygen atoms are identical (homo-nuclear), it is expected that the electrons shared should be equally distributed between the two atoms in the molecule, giving rise to a perfect symmetry. However, since the electrons are in continuous motion, there will be a higher negative charge cloud on one side of the molecule than on the other at any given time. Small poles of charge therefore develop.

This gives rise to instantaneous dipoles, which are temporary, as they keep shifting from side to side. The negative or positive end of one dipole can repel or attract electrons respectively, in a neighbouring molecule, inducing other temporary dipoles. These are called **induced dipoles**.

A force of attraction occurs between the oppositely charged poles of the induced dipoles. This is called an **induced dipole - induced dipole** or **temporary dipole - temporary dipole interaction**.

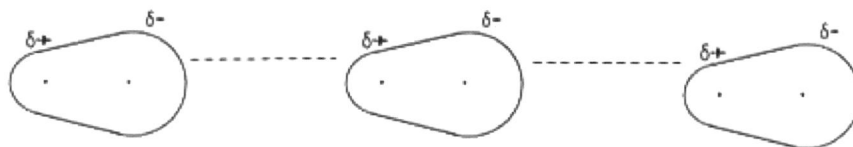

Fig. 5k *Temporary dipole – temporary dipole interaction or induced dipole – induced dipole interaction*

Polarity and dipole moment

When two atoms in a molecule have different electronegativities, there is a separation of charge known as a ***dipole moment***. The greater the difference in electronegativity, the greater is the dipole moment.

		Dipole Moment
H – H		0
	non-polar	
Cl – Cl		0
H – Cl		1.05
	polar	
H – F		greater than 1.05

Table 5.1 *showing dipole moments of polar and non-polar molecules*

In **linear** and **symmetrical** molecules, the dipole moment cancels out, rendering the molecule non-polar.

For example, **CO** is a linear molecule that is **polar**, $\delta+C \overset{x}{\equiv} O\delta-$ where x is the dipole moment.

However, **CO$_2$**, $\delta-O \overset{x}{=} C\delta+ \overset{x}{=} O\delta-$ **non-polar** because the dipole moments x cancel out each other.

NH$_3$, is asymmetrical and **polar** because there is a net dipole moment of x.

CCl$_4$ is symmetrical and **non-polar**, since the dipole moments cancel out each other as shown in the diagram.

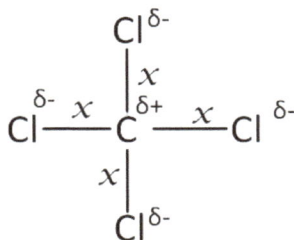

Metallic Bonding

Metallic bonding occurs in metals. It is a force of attraction between a sea of mobile electrons and residual cations. The residual cations are attracted to the sea of electrons. The metallic bond is a strong bond and its strength increases as the number of valence electrons and the charge on the residual cations increase. Consider the metal sodium. The valence electron of each atom is delocalised and forms a sea of negative charge that moves throughout the metal structure as shown in the diagram.

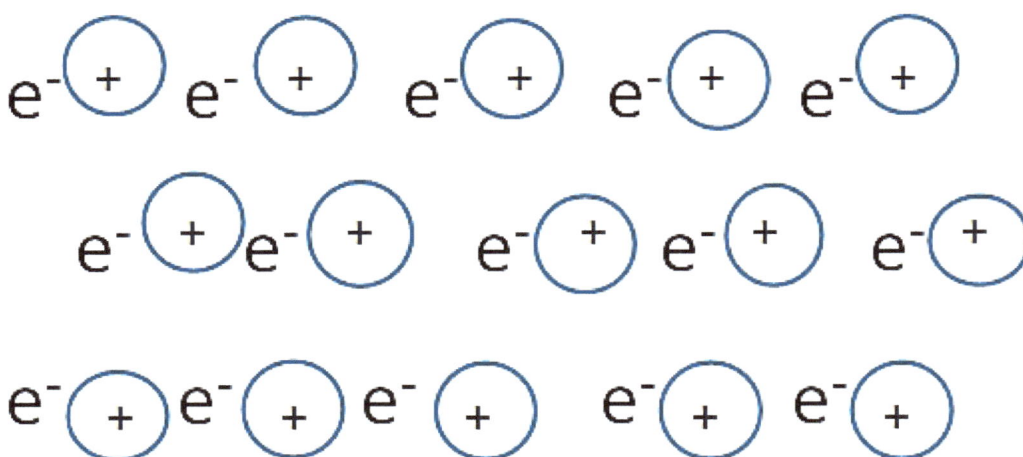

Fig. 5l *Diagram showing metallic bonding in sodium*

Ionic Bonding

When a gaseous atom of sodium metal absorbs enough energy, the valence electron moves into the infinity energy level and no longer experiences the nuclear attraction of the residual cations. The electron is lost and ionisation occurs. A sodium cation is formed. If a gaseous atom of a non-metal gains the electron, it forms an anion. The electrostatic force of attraction between the anion and the cation is called an ***ionic bond*** or an ***electrovalent bond***.

The ionic bond is a strong bond.

The **smaller the ionic radius** is, and the **higher the charge** on the ions, the **stronger** the ionic bond.

For example: **NaCl**   Na 2.8.1   Cl 2.8.7
$1s^22s^22p^63s^1$ $1s^22s^22p^63s^23p^5$

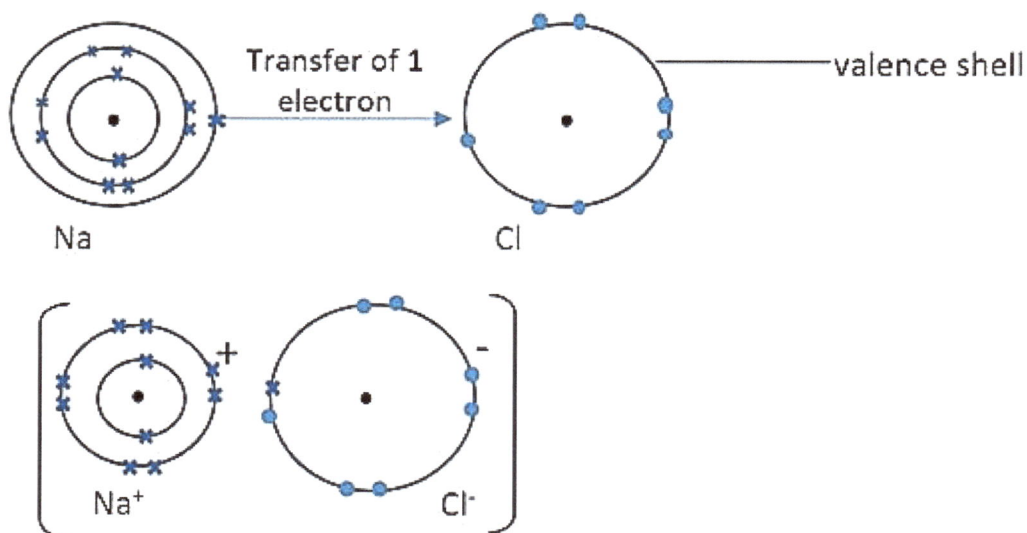

Fig. 5m *showing ionic bonding in NaCl*

Ionic character in covalent bonds

Very few compounds are 100% ionic or 100% covalent.

Ionic character occurs in a covalent molecule or bond if the molecule is polar. The δ^+ and the δ^- introduce some ionic character in the molecule. The greater the difference in electronegativity of the atoms in a molecule is, the greater the dipole moment, and the greater the ionic character.

For example, as the electronegativity difference between the atoms increases from ClF to HF, ionic character increases in the following order:

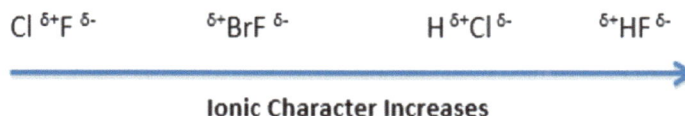

$$Cl\ ^{\delta+}F\ ^{\delta-} \qquad ^{\delta+}BrF\ ^{\delta-} \qquad H\ ^{\delta+}Cl\ ^{\delta-} \qquad ^{\delta+}HF\ ^{\delta-}$$

Ionic Character Increases

Non-polar molecules which have identical atoms, such as H_2 and Cl_2, are purely covalent.

Covalent character in ionic bonds

The following factors will influence covalent character in ionic bonds or ionic compounds:
1. Charge on the cation and anion
2. Size of the cation
3. Size of the anion

Charge on the cation and anion

If the charge on the cation is high, for example Al^{3+}, C^{4+}, it will attract the anion's electrons towards itself, resulting in the distortion of the anion as it shares the anion's electrons. The ability of a cation to attract electrons and distort the anion is called its **polarizing power.** If the anion's charge is high, there is greater repulsion amongst the anion's electrons, making it easier for the cation to attract them and distort the anion.

For example, Al_2O_3 has both covalent and ionic character and is **amphoteric.** $Al^{3+}N^{3-}$ and $Mg^{2+}O^{2-}$ also have ionic and covalent character.

Size of the cation

A small cation will attract an anion's electrons more strongly than a large cation. This results in greater polarisation than a larger cation. This is because the nucleus of a small cation is closer to the anion's electrons and attracts them more strongly. An ionic compound with a small cation would therefore have more covalent character than one with a large cation.

Size of the anion

A larger anion's electrons are attracted less strongly to its own nucleus and can be easily attracted to, and distorted by a small cation.

Therefore, the degree of covalency in an ionic bond is high if:
- The charge on the ions is high
- The cation is small
- The anion is large

A pure ionic compound therefore has a small charge on the cation and also on the anion.
It also has a large cation and a small anion.

The compound with the highest degree of ionic character or the closest to a purely ionic compound is Francium Fluoride. Ionic compounds of group 1 and group 7 elements have the lowest covalent character.

Ceramics

Ceramic materials contain ionic and covalent bonding. These are a group of compounds with overall bonding that is stronger than the bonding in ionic and covalent compounds. Ceramics have very high melting points, and their stability to heat is very high.

They are used to line furnaces, to make floor and roof tiles, crockery and glassware.

They are hard but brittle, and will break easily on impact. This breaking occurs, because a sharp or sudden impact causes like charged ions to slip into layers and repel each other, causing the breakage. This is called **cleavage**.

Exercise 5.3

a. Distinguish between covalent bonding and co-ordinate bonding. Use NH_3 and $AlCl_3$ to illustrate your answer. (6 marks)

b. At 100°C water is in the vapor state. At 25°C it is a liquid. Describe, including illustrations, the forces of attraction that are responsible for the liquefaction of water. (4 marks)

c. i) State the assumptions of the VSEPR Theory. (3 marks)

_____

_____

_____

ii) Use the VSEPR Theory to predict the shapes and bond angles of the PF_3 and H_2S molecules. (4 marks)

d. Consider the following compounds, C_3H_8 and C_3H_7OH. By considering their bonding,

   i)   State which has the higher boiling point (1 mark)_____

   ii) Explain your answer (2 marks)

   _____

   _____

Chapter 6

THE STRUCTURE OF SOLIDS

The particles of solids are arranged in a definite manner. The simplest or smallest unit of a solid crystal, in which the particles are in a definite, geometrical arrangement, is called the **crystal lattice**. These unit cells are repeated throughout the solid crystal.

Lattice units may have the following geometric shapes:

1. Face-centered cubic lattice. For example, NaCl, iodine solid and copper metal.
2. Body-centered cubic lattice. For example, group 1 metals, iron and manganese.
3. Hexagonal lattice. For example, zinc and magnesium metal, graphite.
4. Tetrahedral lattice. For example, ice and diamond.

The main solid structures are:

- **Giant metallic**
- **Simple molecular**
- **Giant ionic**
- **Giant atomic**
- **Giant molecular**

Giant metallic structure

Copper (Cu) has a face centered cubic lattice in which there is a copper atom at the corner of each face of the cube and one in the center of each face. The **co-ordination number** is the number of nearest neighbouring particles such as atoms, ions or molecules in a solid crystal.

In the solid copper crystal, each copper atom is surrounded by 12 nearest, neighbouring copper atoms. Copper is therefore said to have a coordination number of 12. Strong **metallic bonds** hold the copper atoms together in the solid crystal.

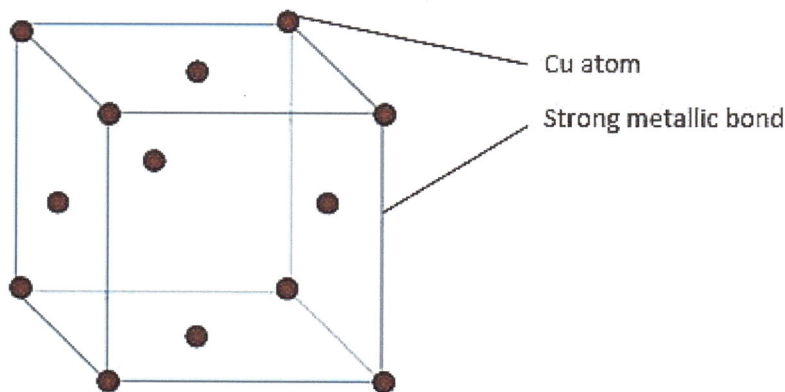

Fig. 6a *Face-centered cubic lattice unit of copper*

Simple molecular structure

Examples of simple molecular solids are solid carbon dioxide (dry ice), solid iodine and ice.

Iodine (I_2) has a face-centered cubic lattice structure with iodine molecules at the corner of each face and in the center of each face.

The molecules are held together by very long, weak, **temporary dipole – temporary dipole** or Van der Waal's forces. Due to the weak Van der Waal's forces, iodine sublimes on heating and has a low density. Because it is non-polar, it is insoluble in water but soluble in most organic solvents.

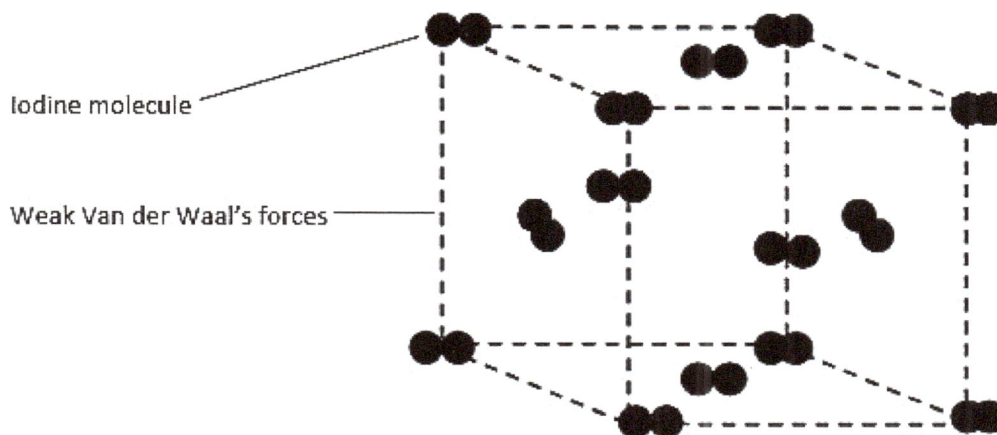

Fig. 6b *Face-centered cubic lattice unit of iodine*

Ice, like iodine, has a simple molecular structure. The water molecules in ice are arranged to give a tetrahedral lattice.
Hydrogen bonds hold the molecules together.

Ice has an open lattice, rather than close packing as in other solids.

Hence, unlike other solids, ice is less dense than its liquid form, water.

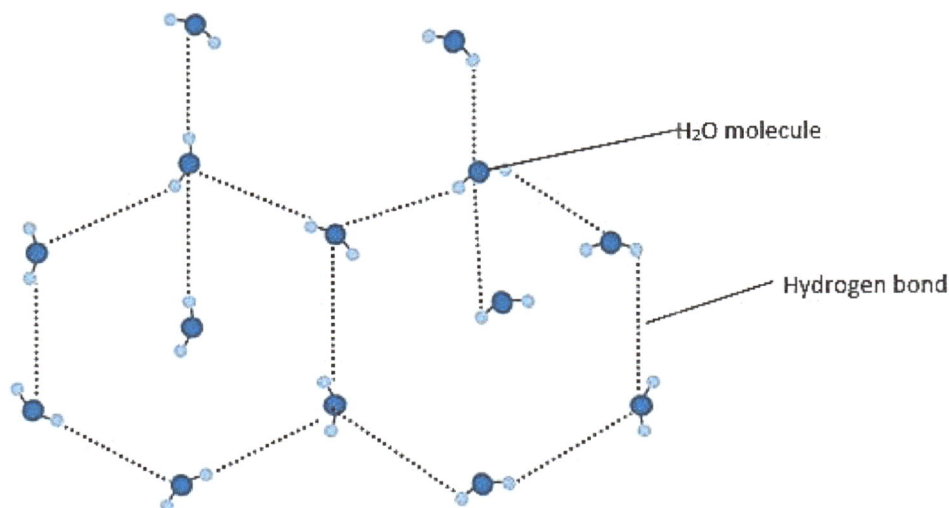

Fig. 6c *the structure of ice*

Giant Ionic Structure

Sodium Chloride (NaCl) has a face-centered cubic crystal lattice, in which there are alternating Na^+ and Cl^- ions on the side of each face and one in the center of each face. Each Na^+ ion is surrounded by 6 nearest neighbouring Cl^- ions and has a coordination number of 6. Each Cl^- ion is surrounded by 6 nearest neighbouring Na^+ ions and has a coordination number of 6. Sodium chloride therefore has a 6:6 coordination number.

Strong **ionic bonds** hold the sodium ions and chloride ions together in the solid crystal.

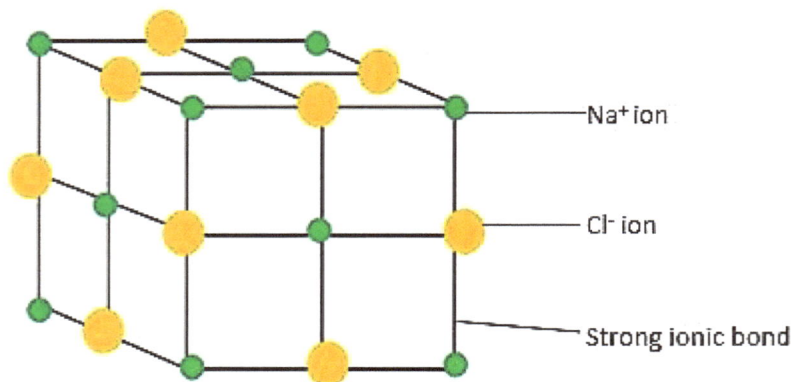

Fig. 6d *Face-centered cubic lattice unit of of NaCl*

Giant Atomic Structure (Macromolecules)

Diamond and graphite are allotropes of carbon which have giant atomic structures. Allotropes are different physical forms of the same element in the same state of matter.

Graphite

The crystal lattice unit of graphite is hexagonal, with each C atom bonded to 3 others by strong **covalent bonds**. The coordination number of carbon in graphite is 3. The hexagonal lattice units are found within layers, which are held together by very weak temporary dipole – temporary dipole interactions or **Van der Waal forces**. This allows the layers to slip and slide over each other. The layers also lower the density of graphite and so graphite is used to make the frames of tennis rackets. It is used to make pencils. It is also used as a lubricant, especially in heavy duty machines.

Only 3 of the 4 electrons are involved in bonding, so the 4th electron of each atom forms a sea of mobile electrons, allowing graphite to be the only non-metal that conducts electricity.

C atom

Strong covalent bond

Weak Van der Waal's force

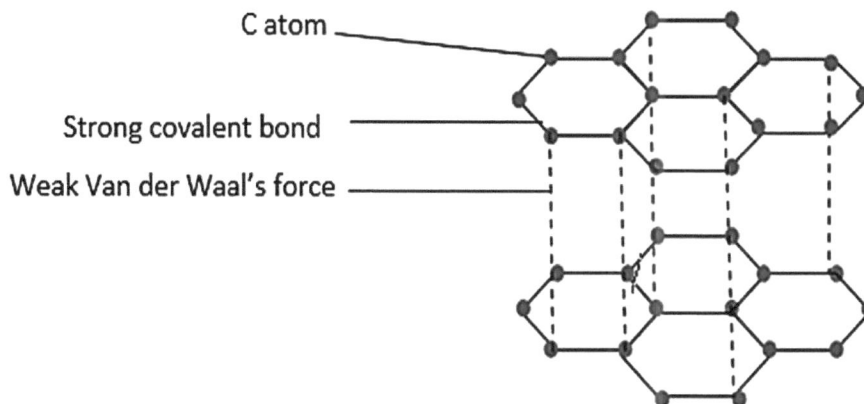

Fig. 6e *Giant atomic structure of graphite*

Diamond

The crystal lattice unit of diamond has a tetrahedral arrangement of carbon atoms. The diamond crystal has a very dense inter-locking network of strong covalent bonds within its structure. This gives diamond a high density and a high melting point. It splits visible light into its constituent coloured wavelengths, allowing diamond to sparkle.

In diamond, each carbon atom is bonded to four others. Therefore, the coordination number of carbon in diamond is 4. Since all 4 valence electrons are involved in bonding, diamond is a non-conductor of electricity.

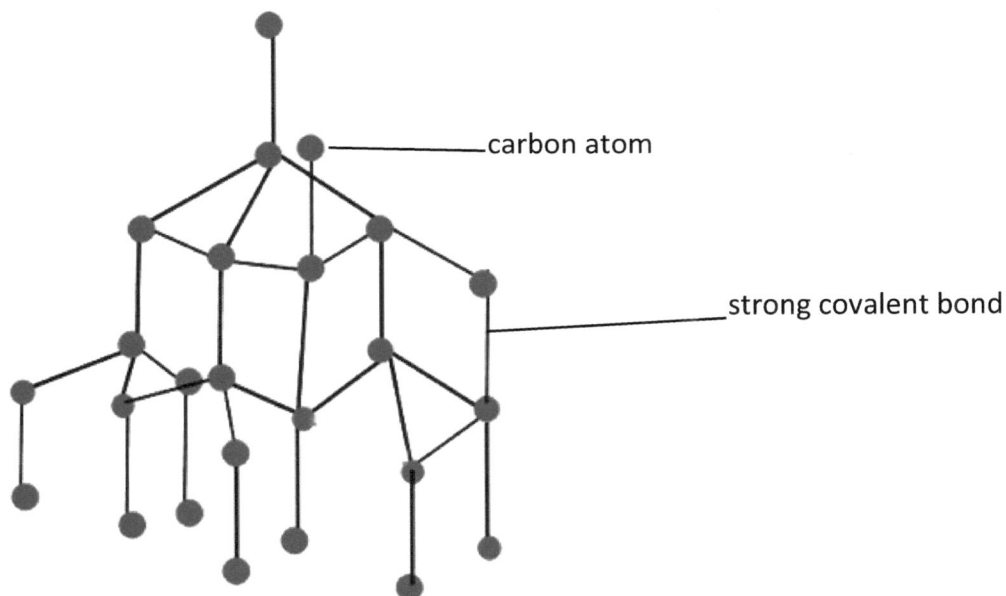

carbon atom

strong covalent bond

Fig.6f *Giant atomic structure of diamond*

Structure	Simple Molecular	Giant Metallic	Giant Ionic	Giant Atomic/Giant Molecular
1. Examples 2. Constituent particles	I_2, S_8, HCl, CH_4, NH_3 Molecules	Na, Fe, Cu Atoms	NaCl, CaO Ions	Diamond, SiC, SiO_2 Atoms
Bonding In the solid	Strong covalent bonds hold atoms together within the separate molecules; separate molecules are held together by weak *Van der Waal's forces*.	Attraction of outer mobile electrons for positive nuclei binds atoms together with strong *metallic bonds*.	Electrostatic attraction of positive ions for negative ions with strong *ionic bonds*.	Atoms are linked through the entire structure by very strong *covalent bonds* from one atom to the other.
Physical Properties 1. Volatility	Volatile	Non-volatile	Non-volatile	Non-volatile
2. Melting and boiling point	Low melting point, low boiling point, low latent heats.	High melting point, high boiling point, high latent heats.	High melting point, high boiling point, high latent heats.	Very high melting point, very high boiling point, very high latent heats.
3. State at room temperature	Usually gases or volatile liquids	Usually solid, except mercury	Solid crystalline	Solid
4. Hardness/ malleability	Soft For example, wax	Hard, yet malleable	Hard and brittle	Very hard and brittle
5. Conductivity (electrical)	Non-conductors when solid and liquid. Polar molecules, For example, HCl and NH_3 dissolve in water to form electrolytes.	Good conductors when solid or liquid. Conductivity is due to the flow of mobile electrons.	Non-conductors when solid. Good conductors when molten or in aqueous solution –because the ions are mobile (electrolytes).	Non-conductors (except graphite). Si and Ge are semi-conductors.
6. Solubility	Non-polar ones are insoluble in polar solvents like H_2O, but soluble in non-polar solvents, like CCl_4.	Insoluble in polar and non-polar solvents, but soluble in liquid metals (alloys).	Soluble in polar solvents, like H_2O. Insoluble in non-polar solvents, like CCl_4.	Insoluble in all solvents.

Table 6.1 *Comparing structure, bonding and physical properties of solid structures*

Examples of **Giant molecular structures** are SiO_2, GeO_2 and SiC.

Structure	Bonding	Examples		M$_r$	B.P/ °C	Solubility
Simple Molecular Structures	temporary dipole – temporary dipole	CH$_4$	No. of e$^-$/ Atomic No. / Bond Strength increases ↓	16	-160	Insoluble in Polar solvents. For example, H$_2$O. Soluble in non-polar solvents.
		SiH$_4$		32	-112.5	
		GeH$_4$		76.6	-90	
Simple Molecular	Permanent dipole – permanent dipole	HCl		37.5	-75	Dissolves in H$_2$O (acidic) Dissolves in non-polar solvents.
	H - Bond	NH$_3$		17	-32	Very soluble in water, less soluble in non-polar solvents.
		HF		10	25	
		H$_2$O		18	100	
Giant metallic	metallic	Mg			649	Insoluble in polar and non-polar solvents. Soluble in liquid metals
		Cu			1090	
Giant Ionic	ionic	NaCl			801	Soluble in water.
		MgBr$_2$			1250	Insoluble in non-polar solvents.
Giant molecular Giant atomic	covalent	SiO$_2$			1610	Insoluble in polar and non-polar solvents.
		C diamond			4827	

Table 6.2 *Comparing Strengths of Bonds using Boiling Point (B.P) Data*

Relating Bonding and Structure to Property of solids

Metallic bond

Metals conduct heat and electricity when solid or molten. This is because the sea of mobile electrons transfer charge throughout the metal. The electrons absorb and reflect light. The reflected light makes metals shiny. The ductility and malleability of metals is related to the shifting of the mobile electrons and their residual cations.

The electrons reflect sound energy causing metals to be sonorous.

Ionic bond

Ionic crystals are brittle. On impact, layers of like-charged ions separate and repel each other in a process called cleavage. This results in breakage. Ionic solids conduct electricity when molten or in aqueous solution as the ions become mobile.

Intermolecular forces of attraction

Solids and liquids with intermolecular forces of attraction have low melting and boiling points and high volatility.

For example, gasoline and kerosene, which are liquids at room temperature, have temporary dipole – temporary dipole forces between non-polar molecules of alkanes. Solids, such as candle wax and iodine also have **temporary dipole – temporary dipole** forces between non-polar molecules. Candle wax melts easily at relatively low temperatures. Other solids such as iodine and carbon dioxide (dry ice), sublime on heating.

The strength of the temporary dipole – temporary dipole force increases with increased number of electrons/ increasing atomic number.

Permanent dipole – permanent dipole forces are stronger than temporary dipole – temporary dipole forces. The greater the electronegativity difference between the atoms in the molecule, the stronger the permanent dipole – permanent dipole force. For example, HBr has weaker forces than HCl.

The Hydrogen-bond
The hydrogen bond is the strongest intermolecular force of attraction. Solid water (ice) and liquid water have hydrogen bonds between the water molecules. The open lattice structure of ice makes it less dense than liquid water and so ice floats on liquid water.

The number of hydrogen bonds per molecule in a hydrogen bonded liquid or solid increases with the number of lone pairs of electrons per molecule. Ammonia has one lone pair and therefore has a lower boiling point than water, which has two lone pairs of electrons. The fluorine atom in the HF molecule has three lone pairs of electrons, and has the strongest hydrogen bonding. However, on heating, dimers are formed at lower temperatures and the boiling point is lower than that of water.

Covalent bond
Solids with covalent bonding such as diamond and silicon (iv) oxide have very high melting and boiling points due to the dense network of strong covalent bonds. They are non-conductors as all their electrons are bonded. Diamond is the hardest substance known, due to the dense network of strong covalent bonds.

Exercise 6.1
a. i) Draw a clearly labeled diagram of a crystal lattice unit of copper. (3 marks)

ii) By considering the fraction of copper atoms shared by adjacent unit cells in solid copper, determine the number of copper atoms that each lattice unit actually contains.(3 marks)

iii) Give three properties of metals and show how they are related to the structure of metals. (6 marks)

_____

_____

iv) Explain the meaning of the term cleavage as it relates to ionic compounds.(2 marks)

_____

_____

_____

b. Consider the molecules CCl_4 and $SiCl_4$.

   i) State their intra-molecular bonding (1 mark) _____

   ii) and their inter-molecular bonding (1 mark)_____

   iii) Explain which of these liquids would have a higher boiling point? (2 marks)
   _____

   iv) Dry ice (solid carbon dioxide) is used on stages for a smoky effect. Explain the principle which results in this effect.(3 marks)

   _____

   _____

Chapter 7
THE MOLE

Definitions

The mole is defined as the amount of a substance that contains as many particles as there are carbon atoms in 12 grams of the C-12 isotope.

1 mole of any substance contains 6×10^{23} particles (atoms, ions or molecules). This is called the Avogadro's number (N).
The mole may be considered the 'currency' of measurements used in chemistry.

Molar mass

This is the mass of 1 mole of an element or compound in grams. Unit = g/mol or gmol^{-1}

Avogadro's Law

Equal volumes of all gases contain the same number of molecules under the same conditions of temperature and pressure. At 25°C and 1 atm pressure (RTP) one mole of any gas occupies a volume of 24 dm^3.

At 0°C, and 1 atm (STP) 1 mole of any gas occupies a volume of 22.4 dm^3.

Empirical Formula

This is the simplest mole ratio in which the atoms of a compound combine. For example CH_2

Molecular Formula

This is the actual number of each atom, in one molecule or one formula unit of a compound. For example, C_3H_6

Relative Masses

These are compared to a standard and have no unit. The C-12 isotope is the standard used.

Relative Atomic Mass (A$_r$)

This is the ratio of the average mass of the atoms of an element to 1/12th the mass of the C-12 isotope.

Relative Molecular Mass/ Relative Formula Mass (M$_r$)

The ratio of the mass of 1 molecule or formula unit of an element or compound to 1/12th the mass of the C-12 isotope.

Calculating Empirical Formula from Combustion Data

Worked example 7.1

a. 0.5438g of a compound containing the elements C, H and O was burnt in pure oxygen. It produced 0.6369g of H_2O and 1.039g of CO_2. Calculate the empirical formula of the compound.

Solution

1 mole of CO_2 has a mass of 44g
44g of CO_2 contains 12g Carbon atoms
∴ *1.039g of CO_2 contains 1,039 ÷44 x 12*
= 0.2834g C

1 mole of H_2O has a mass of 18g
18g of H_2O contains 2g H atoms
0.6369g of H_2O contains 0.6369 ÷18 x 2
=0.0708g H

Mass of O atoms = 0.5438 – (0.2834 + 0.0708)
= 0.1896g O

Atoms	C	H	O
Mass in 0. 5438g of compound.	0.2834	0.0708	0.1896
No. of Moles (÷ A_r)	$\dfrac{0.2834}{12}$ 0.0236	$\dfrac{0.0708}{1}$ 0.0708	$\dfrac{0.1896}{16}$ 0.01185
Mole ratio (÷ Smallest no.)	$\dfrac{0.0236}{0.01185}$ =2.002 = 2	$\dfrac{0.0708}{0.01185}$ =5.97 = 6	$\dfrac{0.01185}{0.01185}$ = 1
Empirical Formula	C_2H_6O		

b. Calculate the molecular formula of the compound given that its formula mass is 138.

Solution

Empirical formula mass = 24+16+6 = 46g
*Molecular formula = 138 ÷ 46 x C_2H_6O = **$C_6H_{18}O_3$***

Molar calculations based on gas volumes

When gases combine they do so in simple mole ratio volumes to each other and to the products, if gaseous.

For example: 2 CO(g)   + O_2 (g)   →   2 CO_2 (g)
Mole ratio=   2 :   1 :   2

Worked example 7.2

From the equation above, calculate the volume of gas produced when 50cm³ of CO is burnt in 20cm³ of O_2.

Solution

	2 CO(g)	+	O_2 (g)	→	2 CO_2 (g)
Mole ratio	2	: 1	:	2
Volume ratio	2	: 1	:	2
Given volume/cm³	50	20		
Reacting/product volumes	40	20	→	40

Answer: 40cm³ CO_2 + 10cm³ of excess CO.

Worked example 7.3

20cm³ of a hydrocarbon gas mixed with 100cm³ of oxygen for combustion. It produces 40cm³ of carbon dioxide and 30cm³ of excess oxygen remains. What is the formula of the hydrocarbon?

Solution

	C_xH_y (g)	+	O_2 (g)	→	$CO_{2(g)} + H_2O_{(l)}$
Given volume/cm³	20	+	100		40
Reacting volume/cm³	20	+	(100-30)	→	40
Volume ratio	2	:	7	:	4
Mole ratio	2	:	7	:	4
Equation =	$2C_xH_y$	+	$7O_2 →$		$4CO_2 + H_2O$
Balancing the O, atoms	$2C_xH_y$	+	$7O_2 →$		$4CO_2 + 6 H_2O$

Balancing the C and H atoms, x=2 and y=6

Answer ∴ Molecular formula = C_2H_6

Worked example 7.4

a) Calculate the mass of $MgCl_2$ produced when 19.0g of Mg is reacted with dilute HCl solution.

b) What volume of gas is produced at S.T.P?

Solution

	$Mg_{(s)}$	+	$2HCl_{(aq)}$ →		$MgCl_{2 (aq)} + H_{2(g)}$
No. of moles	1		2		1 1
Molar mass/vol.	24g				24+71=95g 22.4dm³ (S.T.P)

a) From the equation 24g Mg produced 95 $MgCl_2$

∴ 19g Mg produced 95 ÷24 x 19 = **75.21g $MgCl_2$** *(Answer)*

b)   24g Mg produced 22.4 dm³ H_2 at S.T.P

∴ 19g Mg produced 22.4 ÷ 24 x 19 = **17.73dm³ (Answer)**

Worked example 7.5
An average of 15.60cm³ H_2O_2 required 25.0cm³ of 0.10moldm⁻³ acidified $KMnO_4$ for oxidation. Calculate:

a) The molar concentration of the H_2O_2
b) The mass concentration of H_2O_2

Solution
a) 1000cm³ $KMnO_4$ contains 0.10 mol MnO_4^-/H^+ ions
∴ 25.0cm³ $KMnO_4$ contains 0.1 ÷1000 x 25.0 = **0.0025** mol .

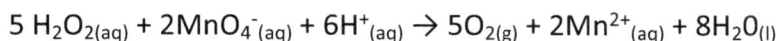

$5 H_2O_{2(aq)} + 2MnO_4^-_{(aq)} + 6H^+_{(aq)} \rightarrow 5O_{2(g)} + 2Mn^{2+}_{(aq)} + 8H_2O_{(l)}$

From the equation 2 mol MnO_4^- oxidise 5 mol x 25.0 H_2O_2
Therefore 0.0025mol MnO_4^- oxidise 5 ÷2 x 0.0025 = **0.00625** mol H_2O_2
15.60 cm³ H_2O_2 contains 0.00625 mol.
∴ 1000cm³ H_2O_2 contains 0.00625 ÷15.60 x 1000 x = **0.400 moldm⁻³** (*Answer*)

b) 1 mol H_2O_2 has a mass of (2 + 32) **= 34g**

∴ 0.400 mol has a mass of 34 x 0.400 = **13.62g dm⁻³** (*Answer*)

Note – *Construction of appropriate statements for calculations on the mole is imperative.*

Exercise 7.1
a. 10cm³ of a gaseous hydrocarbon was mixed with 80cm³ of O_2 and sparked. After sparking the mixture, the volume of gases remaining was 60cm³. On adding NaOH solution, the volume decreased to 20cm³. All measurements were made at RTP

i) Explain the role of the sodium hydroxide solution. (1 mark)

_____

ii) Calculate the molecular formula of the hydrocarbon. (3 marks)

iii) Calculate the volume of gas produced when 60cm³ of nitrogen and 200cm³ of hydrogen are reacted together. (3 marks)

b. Define the term mole. (1mark)

_____

_____

c. Distinguish between the terms *Empirical formula* and *Molecular Formula.* (2 marks)

_____

_____

_____

d. 1.0g of a compound A containing only carbon, hydrogen and oxygen, was completely combusted. It produced 2.3g of CO_2 and 0.93g of water. The relative molecular mass of compound A is 116. Calculate the empirical and molecular formulae of A. (5 marks)

Exercise 7.2

a. Calculate the composition by volume of the resulting gas when 200cm³ of nitrogen is reacted with 500cm³ of hydrogen. (2 marks)

b. What mass of solid is precipitated when excess lead (II) nitrate is added to 10cm³ of 0.5 moldm⁻³ potassium iodide solution? (3 marks)

c. What volume at S.T.P of ammonia is produced when 12g of hydrogen react completely with nitrogen? (3 marks)

Chapter 8
OXIDATION and REDUCTION (REDOX)

Redox means reduction and oxidation. Oxidation and reduction are complementary processes. This means that one does not occur without the other.

During a redox reaction, the reactant that is oxidised acts as the reducing agent and that which is reduced as the oxidising agent.

Definitions of oxidation and reduction

Oxidation and reduction can be defined in terms of the gain and loss of oxygen and hydrogen atoms, the gain and loss of electrons and the increase and decrease in oxidation number of an atom.

Oxidation (oxidised)	Reduction (reduced)
The gain of O atoms	The loss of O atoms
The loss of H atoms	The gain of H atoms
The loss of electrons	The gain of electrons
An increase in oxidation number/state of an atom	A decrease in oxidation number/state of an atom

Table 8.1 *Definitions showing the complementary relationship between oxidation and reduction*

Based on the table definitions, species that are oxidised are reducing agents. **Reducing agents**, therefore, accept oxygen atoms, donate hydrogen atoms and electrons and they undergo an increase in the oxidation state.

An oxidising agent donates oxygen atoms, accepts hydrogen atoms and electrons and undergoes a decrease in oxidation state.

What is the meaning of oxidation state or oxidation number?

This gives the combining power of an atom in a compound. The combining power of C in CO and in CO_2 is different. In CO, C combines with one oxygen atom, but combines with two oxygen atoms in CO_2. Elements have an oxidation number of zero, since they are not combined.

Common Oxidation States

Elements of Group 1 have a +1 oxidation state. Group 2, +2, Group 3, +3 and Group 7, -1.

The oxidation state of the oxygen atom, O is, -2 except in peroxides such as H_2O_2 where it is -1 and the hydrogen atom, H is, + 1 except in hydrides of active metals such as NaH where it is -1.

Calculating oxidation numbers/states

The algebraic sum of the oxidation numbers of the atoms in an ion or compound is equal to the charge on the ion or the compound.

Worked example 8.1

Calculate the oxidation state of

a) The Mn atom in each of the following:

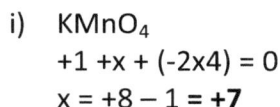

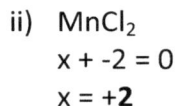

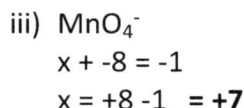

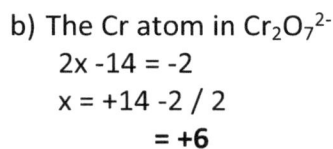

    i) $KMnO_4$

        $+1 +x + (-2 \times 4) = 0$

        $x = +8 - 1 = \mathbf{+7}$

    ii) $MnCl_2$

        $x + -2 = 0$

        $x = \mathbf{+2}$

    iii) MnO_4^-

        $x + -8 = -1$

        $x = +8 -1 = \mathbf{+7}$

b) The Cr atom in $Cr_2O_7^{2-}$

    $2x -14 = -2$

    $x = +14 -2 / 2$

          $= \mathbf{+6}$

Balancing Redox Equations

Worked example 8.2

Write a balanced equation for:

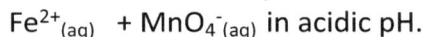

$Fe^{2+}_{(aq)} + MnO_4^-{}_{(aq)}$ in acidic pH.

In balancing redox equations, the oxidation and the reduction half reactions must be separately balanced and then added together at the end to give the fully balanced equation.

Step 1 - *Balance the atoms*.

If a reaction is carried out in acid pH, introduce H^+ ions and H_2O on opposite sides of the equation. If it is carried out in basic pH, introduce OH^- ions and H_2O on opposite sides of the equation.

Solution

MnO_4^- $+$ $8H^+ \rightarrow$ $Mn^{2+} + 4H_2O$

Purple   colourless

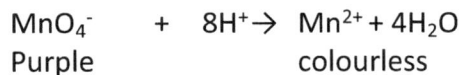

Step 2 - Balance the charge by adding or removing electrons from the reactant side.

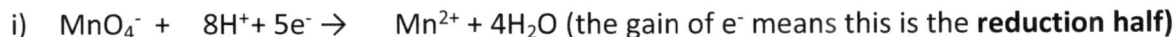

i) $MnO_4^- + $ $8H^+ + 5e^- \rightarrow$ $Mn^{2+} + 4H_2O$ (the gain of e^- means this is the **reduction half**)

Repeat steps 1 and 2 for the other half reaction

ii) $Fe^{2+} - e^-$ $\rightarrow$ Fe^{3+}(loss of electrons **= oxidation half**)

Step 3 - Equalize the number of electrons lost and gained by multiplying by a small, appropriate whole number.

Multiply *(ii)* x 5

$5\,Fe^{2+} - 5e^- \rightarrow 5Fe^{3+}$

Step 4 - Add the two half equations.

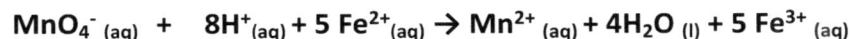

$MnO_4^-{}_{(aq)} + $ $8H^+_{(aq)} + 5\,Fe^{2+}_{(aq)} \rightarrow Mn^{2+}{}_{(aq)} + 4H_2O_{(l)} + 5\,Fe^{3+}_{(aq)}$

Worked example 8.

Balance the equation for the reaction, $SO_3^{2-} + MnO_4^- \rightarrow SO_4^{2-} + MnO_2$ in basic pH.

Note. *You can place the OH^-/H^+ and the H_2O on any side of the equation.*

$$SO_3^{2-} + 2OH^- \rightarrow SO_4^{2-} + H_2O$$

i) $SO_3^{2-} + 2OH^- -2e^- \rightarrow SO_4^{2-} + H_2O$ (oxidation half)

$$MnO_4^- + 2H_2O \rightarrow MnO_2 + 4OH^-$$

ii) $MnO_4^- + 2H_2O + 3e^- \rightarrow MnO_2 + 4OH^-$ (reduction half)

Multiply (i) by 3: $\quad 3SO_3^{2-} + 6OH^- - 6e^- \rightarrow 3SO_4^{2-} + 3H_2O$

Multiply (ii) by 2: $\quad 2MnO_4^- + 4H_2O + 6e^- \rightarrow 2MnO_2 + 8OH^-$

Add both halves: $\quad 3SO_3^{2-}{}_{(aq)} + 2MnO_4^-{}_{(aq)} + H_2O_{(l)} \rightarrow 3SO_4^{2-}{}_{(aq)} + 2MnO_{2(s)} + 2OH^-{}_{(aq)}$

$\qquad\qquad\qquad +4 \qquad\qquad\quad +7 \qquad\qquad\qquad\qquad +6 \qquad\qquad\quad +4$

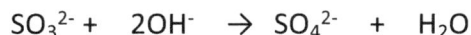

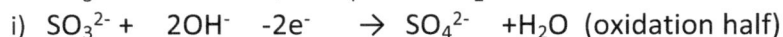

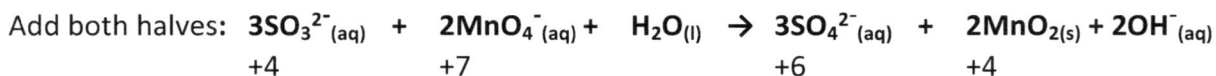

By considering the changes in oxidation numbers, the oxidising agent and reducing agent in the equation above can be identified as follows:

Assign the oxidation numbers to only those atoms that have undergone a change in oxidation state as above.

The Mn atom in MnO_4^- has undergone a **decrease** in its oxidation number from +7 to +4 in MnO_2.

Therefore, MnO_4^- is **reduced** and it is the **oxidising agent**.

The S atom in SO_3^{2-} has undergone an **increase** in its oxidation number from +4 to +6 in SO_4^{2-}.

Therefore, SO_3^{2-} is **oxidised** and it is the **reducing agent**.

Some examples of **oxidising agents** are non metals in general, O_2, F_2, Cl_2, Br_2, H_2O_2, $MnO_4^-/H+$, $Cr_2O_7^{2-}/H^+$, conc. H_2SO_4.

Some examples of **reducing agents** are metals, H_2, H_2S, ethanol, NH_3, Fe^{2+}, I^-, halide ions, H_2O_2.

A breathalyser test for the detection of alcohol on the breath of intoxicated drivers is based on the reducing properties of alcohols, specifically ethanol. The instrument uses the oxidising agent potassium dichromate (VI). The ethanol reduces the dichromate ion from its +6 oxidation state to the +3 oxidation state in Cr^{3+}. A colour change from orange to green is observed. The instrument is calibrated in such a way that the intensity of the green determines the meter reading and indicates the percentage of alcohol in the breath. More recent breathalyser tests use electrolysis and infrared radiation.

Tests for oxidising and reducing agents

To test an unknown substance to determine if it is an oxidising agent, it must be reacted with a reducing agent that will produce a very visible colour change. Similarly, if an unknown substance is to be tested to determine if it is a reducing agent, it should be reacted with an oxidising agent that will produce a visible colour change.

Test	Observation	Inference
(a) Oxidising agent		
To about 2 cm³ of "A," add a few drops of the reducing agent Fe^{2+} ion, followed by $OH^-_{(aq)}$ For example, NaOH **OR** Add a few drops of KI(aq) to A	Colourless to pale yellow Rust brown precipitate Colourless solution turns yellow or brown	Fe^{2+} oxidised by A to Fe^{3+} ion, $Fe^{2+}_{aq} - 1e- \rightarrow Fe^{3+}_{aq}$ Fe^{3+} precipitates $Fe(OH)_3$ with $OH^-_{(aq)}$ $Fe^{3+}_{(aq)} + 3OH^-_{(aq)} \rightarrow Fe(OH)_{3\,(s)}$ I^- is oxidised by A to I_2 $2I^-_{(aq)} - 2e^- \rightarrow I_{2(s)}$ **A is therefore an oxidising agent**
(b) Reducing Agent		
To about 2 cm³ of "B" add a few drops of the oxidising agent For example, $MnO_4^-{}_{(aq)} / H^+_{(aq)}$ **OR** Add a few drops of $Cr_2O_7^{2-}{}_{(aq)}/H^+_{(aq)}$	Purple to colourless Orange to green	MnO_4^- reduced to Mn^{2+} by B $MnO_4^- + 5e^- + 8H^+ \rightarrow Mn^{2+} + 4H_2O$ $Cr_2O_7^{2-}$ reduced to Cr^{3+} **B is therefore a reducing agent** $Cr_2O_7^{2-} + 6e^- + 14 H^+ \rightarrow 2Cr^{3+} + 7 H_2O$

Table 8.1 *Tests for oxidising and reducing agents*

The Displacement reactions of oxidising and reducing agents

A strong oxidising agent will displace a weaker one from its aqueous ions. Similarly, a strong reducing agent will displace a weaker one from its aqueous ions.

In Group 7, fluorine (F_2) is the strongest oxidising agent followed Cl_2, Br_2 and I_2. Chlorine, when reacted with a colourless aqueous solution of Br^- ions, produces a reddish brown solution of bromine (Br_2). The halogens are more soluble in organic solvents such as hexane, than in water and their colours are much more apparent.

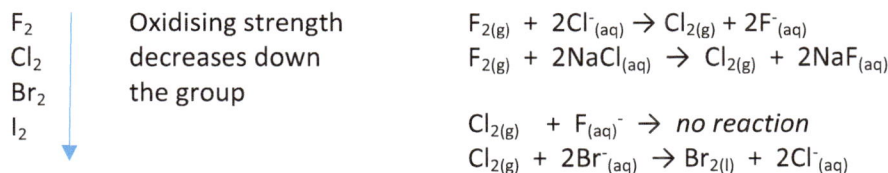

F_2 Oxidising strength $F_{2(g)} + 2Cl^-_{(aq)} \rightarrow Cl_{2(g)} + 2F^-_{(aq)}$
Cl_2 decreases down $F_{2(g)} + 2NaCl_{(aq)} \rightarrow Cl_{2(g)} + 2NaF_{(aq)}$
Br_2 the group
I_2 $Cl_{2(g)} + F^-_{(aq)} \rightarrow$ *no reaction*
                               $Cl_{2(g)} + 2Br^-_{(aq)} \rightarrow Br_{2(l)} + 2Cl^-_{(aq)}$

F_2 will displace all of the halogens below it from their aqueous solutions. I_2 will displace none of the halogens above it, but will displace astatine, which lies below it.

Metals are reducing agents. Magnesium (Mg) is a stronger reducing agent than Zinc (Zn) and will therefore displace zinc from its aqueous ions.
Metals in decreasing reducing strength – Mg> Zn> Cu> Ag
For example, $Mg_{(s)} + Zn^{2+}_{(aq)} \rightarrow Zn_{(s)} + Mg^{2+}_{(aq)}$
$Mg_{(s)} + Zn(NO_3)_{2(aq)} \rightarrow Zn_{(s)} + Mg(NO_3)_{2(s)}$

$Zn_{(s)} + Cu^{2+}_{(aq)} \rightarrow Cu_{(s)} + Zn^{2+}_{(aq)}$
        Blue   pink-brown/ peach   colourless

When a reactant is both oxidised and reduced in the same reaction it is called a **disproportionation reaction**.

Exercise 8.1

a. Explain the meaning of the term 'Redox'. (2 marks)

_____

_____

b. By considering the following redox equation:

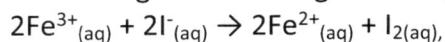

$$2Fe^{3+}_{(aq)} + 2I^-_{(aq)} \rightarrow 2Fe^{2+}_{(aq)} + I_{2(aq)},$$

Use changes in oxidation states to explain:
i) What has been oxidised and what has been reduced. (3 marks)

_____

_____

ii) What is the oxidising agent and what is the reducing agent. (2 marks)

_____

_____

c. Describe what will be observed when the following reactions are carried out:
i) An acidified solution of $KMnO_{4(aq)}$ is added drop-wise to a solution of iron(II) sulphate. (2 marks)

_____

ii) An aqueous solution of chlorine is added drop-wise to an aqueous solution of sodium bromide. (2 marks)

_____

iii) Iron filings are added to a solution of copper (II) sulphate. (2 marks)

_____

8.2 Write balanced redox equations for the following reactions:

a. $MnO_2 + SO_3^{2-} \rightarrow Mn^{2+} + S_2O_6^{2-}$, in acid pH   (4 marks)

b. $HXeO_4^- \rightarrow XeO_6^{4-} + Xe + O_2$, in basic pH (4 marks)

c. Consider: $MnO_4^-{}_{(aq)} + 5Fe^{2+}{}_{(aq)} + 8H^+{}_{(aq)} \rightarrow Mn^{2+}{}_{(aq)} + 5Fe^{3+}{}_{(aq)} + 4H_2O_{(l)}$
Deduce the oxidation half reaction and the reduction half reaction for the above reaction.
(2 marks)

d. When a reactant in a reaction undergoes both oxidation and reduction the redox reaction is called a disproportionation reaction.

By considering the following equation: $Cl_{2(g)} + 2NaOH_{(aq)} \rightarrow NaCl_{(aq)} + NaOCl_{(aq)}$, use oxidation numbers to explain if the above reaction is a disproportionation reaction. (2 marks)

Chapter 9

Energetics pertains to the energy changes that occur during the course of a chemical reaction. During chemical reactions old bonds are broken and new bonds are made.

When bonds are broken, energy is used/ absorbed but when bonds are made, energy is released or given off.

Enthalpy

All matter has an energy content called **enthalpy (H)**. In a chemical reaction, the reactants and products constitute the **chemical system**. Everything that is outside of these, make up the **surroundings** of the system, for example the glassware and the external atmosphere.

The **net enthalpy change** of a chemical reaction is the sum of the energy absorbed and released during the course of that reaction. If there is a net absorption of energy, then the reaction system is said to be **endothermic** and the temperature of the surroundings falls.

However, if there is a net release of energy into the surroundings the reaction is said to be **exothermic** and the temperature of the surroundings rises.

The enthalpy change of a reaction can be calculated from: $\Delta H = H_{products} - H_{reactants}$

For example, if **hypothetically**, the energy values are 100 units for the reactant and 120 for the product as shown below.

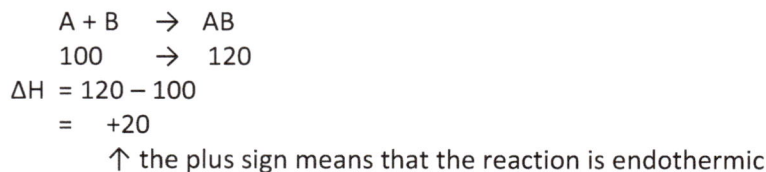

$$A + B \rightarrow AB$$
$$100 \rightarrow 120$$
$$\Delta H = 120 - 100$$
$$= +20$$

↑ the plus sign means that the reaction is endothermic

However, if the energy values are 100 for the reactant and 60 for the product as shown below:

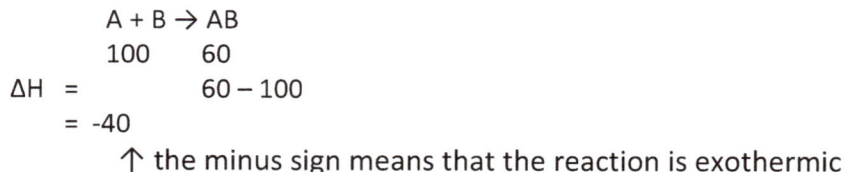

$$A + B \rightarrow AB$$
$$100 \quad 60$$
$$\Delta H = \quad 60 - 100$$
$$= -40$$

↑ the minus sign means that the reaction is exothermic

Definitions of Standard enthalpy changes of reactions ($\Delta H^\ominus$)

Standard enthalpy change of reaction ($\Delta H^\ominus_{reaction}$)

This is the heat/energy absorbed or released during a chemical reaction, by molar quantities of reactants in accordance with the balanced equation at 298K (25°C) and 1 atm pressure, (standard conditions).

$\Delta H^{\square}$ Atomisation/ standard heat of atomisation ($\Delta H^{\square}_{at}$)

Energy absorbed/required to form 1 mole of gaseous atoms from its element in its standard state at 25°C and 1 atm pressure (standard conditions.)

$$\frac{1}{2}Cl_{2(g)} \rightarrow Cl_{(g)}$$
$$Na_{(s)} \rightarrow Na_{(g)}$$

$\Delta H^{\square}$ Formation/ standard heat of formation ($\Delta H^{\square}_{f}$)

Energy absorbed or released when 1 mole of a compound is formed from its elements in their standard state at 25°C and 1 atm.

$$Na_{(s)} + \frac{1}{2}Cl_{2(g)} \rightarrow NaCl_{(s)}$$
$$2C_{(s)} + 3H_{2(g)} \rightarrow C_2H_{6(g)}$$

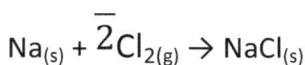

$\Delta H^{\square}$ Combustion / standard heat of combustion $\Delta H^{\square}_{c}$

Energy released when 1 mole of an element or compound is completely burnt in oxygen at 25°C and 1 atm.

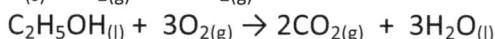

$$C_{(s)} + O_{2(g)} \rightarrow CO_{2(g)}$$
$$C_2H_5OH_{(l)} + 3O_{2(g)} \rightarrow 2CO_{2(g)} + 3H_2O_{(l)}$$

$\Delta H^{\square}$ Neutralisation/ standard heat of neutralisation $\Delta H^{\square}_{n}$

Energy released when 1 mole of H_2O is formed from the $H^+_{(aq)}$ and $OH^-_{(aq)}$ ions at 25°C and 1 atm. (standard conditions)

$$H^+_{(aq)} + OH^-_{(aq)} \rightarrow H_2O_{(l)}$$

$\Delta H^{\square}$ Ionisation / Ionisation energy($\Delta H^{\square}_{i}$)

This is the energy required to form 1 mole of gaseous cations by the loss of 1 mole of electrons at 25°C and 1 atm.

$\Delta H^{\square}$ Electron Affinity($\Delta H^{\square}_{ea}$)

Energy absorbed or released when I mole of gaseous anions are formed by the gain of 1 mole of electrons at standard conditions.

$$Cl_{(g)} + 1e^- \rightarrow Cl^-_{(g)}$$

The first electron affinity is always exothermic however second and successive electron affinities can be either exothermic or endothermic.

$\Delta H^{\square}_{latt}$ /Lattice Energy $\Delta H^{\square}_{latt}$

Energy released when 1 mole of an ionic compound is formed from its gaseous ions at standard conditions.

$$Na^+_{(g)} + Cl^-_{(g)} \rightarrow NaCl_{(s)}$$

$\Delta H^{\square}$bond /Bond Energy ($\Delta H^{\square}_{be}$)

Energy absorbed or released when 1 mole of covalent bond is broken or formed at 25°C and 1 atm.

$$N \equiv N_{(g)} \rightarrow 2N_{(g)}$$

$\Delta H^{\square}$hydration/ Hydration Energy ($\Delta H^{\square}_{hyd}$)

Energy released when 1 mole of gaseous ions are surrounded by H_2O molecules to form an infinitely dilute solution at 25°C and 1 atm.

$\Delta H^{\ominus}$solution ($\Delta H^{\ominus}_{soln}$)

Energy absorbed or released when 1 mole of a solute is complete dissolved in a solvent (usually water) to form a solution of infinite dilution at standard temperature and pressure of 25°C and 1 atm.

Energy Diagrams

Energy diagrams describe energy changes occurring during chemical reactions in a graphical way. There are three types of energy diagrams.

1. Energy level diagrams
2. Energy cycle diagrams
3. Energy profile diagrams

An energy level diagram shows the energy content of the reactants and the products in a reaction.

Energy Level diagrams

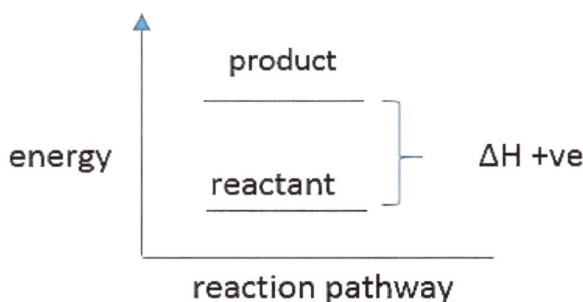

Fig. 9a *Energy level diagram for an endothermic reaction*

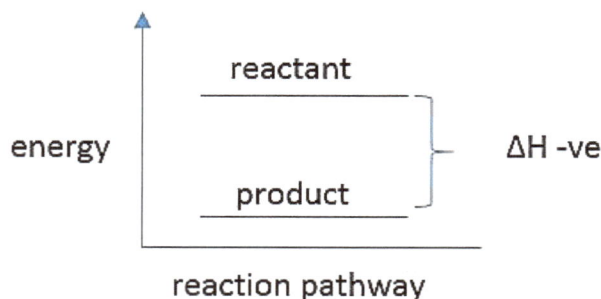

Fig. 9b *Energy level diagram for an exothermic reaction*

Energy Cycle Diagrams

An energy cycle diagram describes and shows all the energy changes that occur during the course of the chemical reaction. The following is an energy cycle diagram for the formation of NaCl.

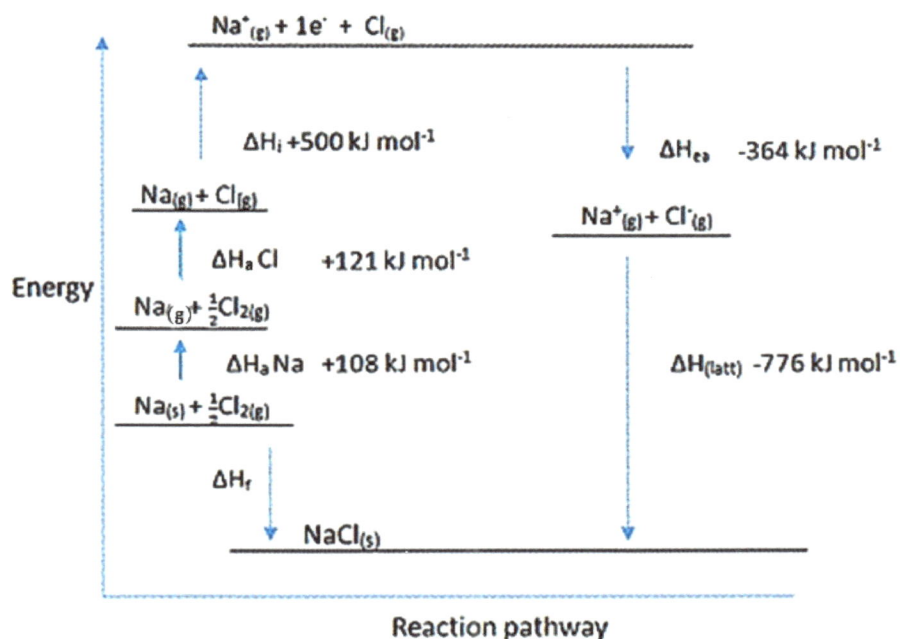

Fig. 9c *Energy cycle diagram/ Born-Haber cycle for the formation of sodium chloride*

Worked example 9.1

Calculate the enthalpy change of formation of NaCl.

ΔH_f is equal to the sum of all the energy changes during the course of the formation of the compound.

ΔH_f $= \sum \Delta H_a Na + \Delta H_a Cl + \Delta H_i + \Delta H_{ea} + \Delta H_{latt}$

$= 108 + 121 + 500 - 364 - 776$

Answer $= -411 kJ\ mol^{-1}$

If the enthalpy change of formation is known and the lattice energy unknown, the lattice energy can be calculated as follows:

ΔH_f $= \sum \Delta H_a Na + \Delta H_a Cl + \Delta H_i + \Delta H_{ea} + \Delta H_{latt}$

-411 $= 108 + 121 + 500 - 364 + \Delta H_{latt}$

-411 $= 365 + \Delta H_{latt}$

$-411 - 365 = \Delta H_{latt}$

Answer $-776\ kJ\ mol^{-1} = \Delta H_{latt}$

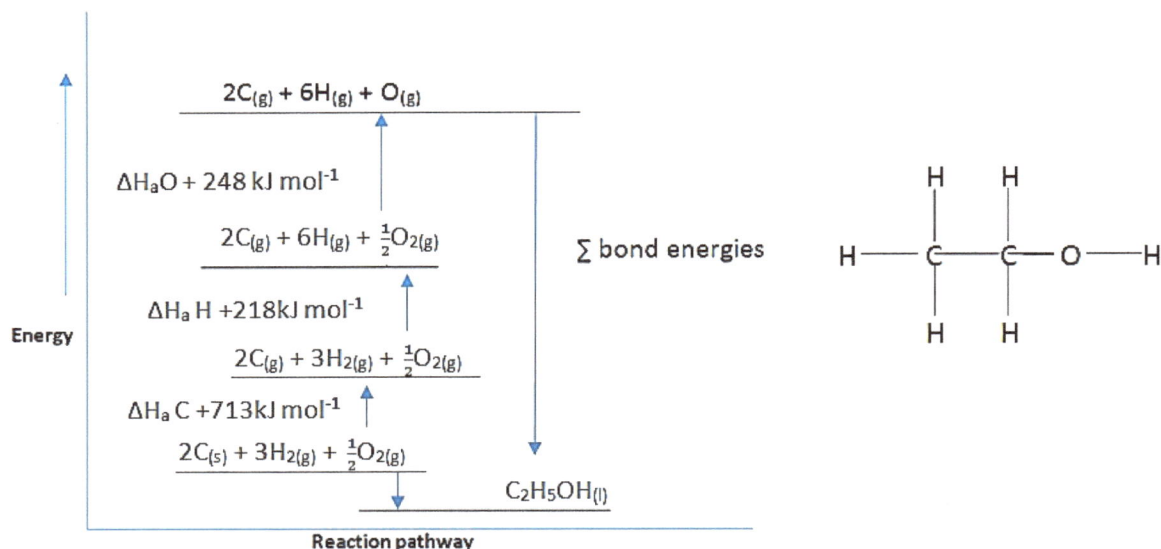

Fig. 9d *Energy cycle diagram for the formation of ethanol*

Worked example 9.2

Use the atomization energies and bond energies to calculate ΔH_f for ethanol.

ΔH_f = sum of atomization energies + sum of bond energies.

Total atomization energies 2x $\Delta H_a C$ + 6x $\Delta H_a H$ + 1x $\Delta H_a O$

$$= (2 \times +713) + (6 \times +218) + (1 \times +248)$$

$$= +1426kJ + 1308kJ + 248kJ$$

$$= +2982kJ$$

Total Bond Energy released =

5, C – H	5 x -410 = -2050kJ	
1, C – C	1 x -350 = -350kJ	
1, C – O	1 x -360 = -360kJ	
1, O – H	1 x -460 = -460kJ	

= -2050 - 350 - 360 - 460= **-3220 kJ**

$\therefore$ ΔH_f = 2982 + (-3220)

Answer = **-238kJ mol^{-1} of ethanol**

Hess's Law of Constant Heat Summation

Hess's Law states that the enthalpy change in a reaction is the same regardless of the pathway by which the reaction occurs, provided that the initial and final conditions of the pathways are the same.

Hess's Law is applied in calculating the enthalpy change of reactions in which it is difficult to measure the enthalpy change directly in the laboratory.

Another way of defining Hess's Law of constant heat summation is: if a reaction can be carried out in an alternate route with steps, the enthalpy change of the reaction is the sum of the enthalpy changes of the steps in the alternate route, provided that the initial and final conditions remain the same. The ΔH of the reaction is the same regardless of the route taken.

Worked example 9.3

Use the reactions (i) and (ii) below to calculate the enthalpy change of formation of Fe_2O_3

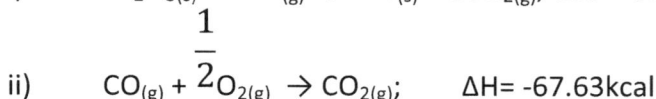

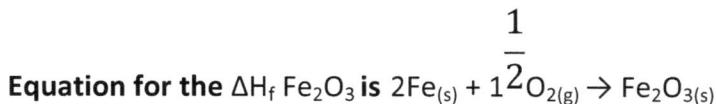

i) $Fe_2O_{3(s)} + 3CO_{(g)} \rightarrow 2Fe_{(s)} + 3CO_{2(g)}$; ΔH= -6.39kcal

ii) $CO_{(g)} + \frac{1}{2}O_{2(g)} \rightarrow CO_{2(g)}$; ΔH= -67.63kcal

Equation for the ΔH_f Fe_2O_3 **is** $2Fe_{(s)} + 1\frac{1}{2}O_{2(g)} \rightarrow Fe_2O_{3(s)}$

Calculation

Step 1 - *Reverse i*: $2Fe_{(s)} + 3CO_{2(g)} \rightarrow$ $Fe_2O_{3(s)} + 3CO_{(g)}$ ΔH= +6.39kcal (the sign is reversed) (iii)

Step 2 *Multiply ii by 3*: $3CO_{(g)} + \frac{3}{2}O_{2(g)} \rightarrow 3CO_{2(g)}$ ΔH= 3 x -67.63 = -202.89kcal (iv)

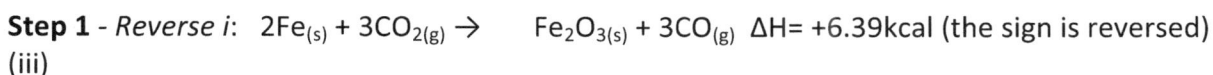

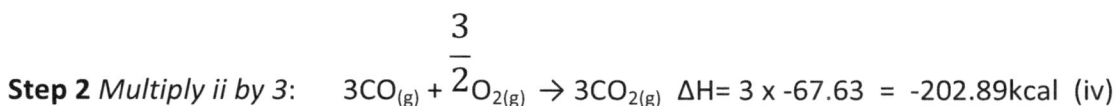

Step 3 *Apply Hess's law of heat summation by adding equations (iii) and (iv).*

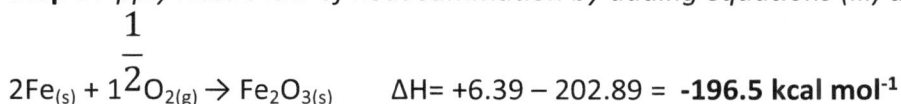

$2Fe_{(s)} + 1\frac{1}{2}O_{2(g)} \rightarrow Fe_2O_{3(s)}$ ΔH= +6.39 – 202.89 = **-196.5 kcal mol^{-1}**

Steps in the experimental determination of Enthalpy change of Reaction.

- Use a known volume of water for dissolving solutes and
- Known volumes of aqueous solutions of reactants.
- Use a known mass or known number of moles of solid reactant.
- Record temperature of liquids before and after mixing reactants, using a thermometer.
- Carry out experiment in a plastic/styrofoam cup to avoid heat exchange with the surroundings.

Analysis of data

- Calculate the difference between the initial and final temperatures, ΔT
- Calculate the energy released or absorbed, ΔE, by the number of moles of reactants used during the reaction using: $\Delta E = m\ c\ \Delta T$, where m = mass of water or aqueous solution. (Density of water is 1 g/cm^3), c = the specific heat capacity of water.
- Calculate the energy absorbed or released by 1 mole of the reactant or product formed.

Bond Energies

Bond energies indicate the strength of covalent bonds. The higher the bond energy is, the stronger the covalent bond.

Bond energies can therefore be used to compare the strength of covalent bonds. This list of bond energies shows that the $N \equiv N$ bond is very strong as compared to the C-C bond. This high bond energy for N_2 explains why nitrogen does not react under normal laboratory conditions. The bond, however, can be broken by lightning or by the heat generated in internal combustion engines. Under these conditions, oxides of nitrogen are formed.

Examples of Bond Energies
C - C = 350 KJmol^{-1}
C – H = 410 KJmol^{-1}
C – O = 360 KJmol^{-1}
O – H = 460 KJmol^{-1}
N $\equiv$ N = 945 KJmol^{-1}

Bond energies are used in calculating enthalpy changes of reactions. They give an understanding of the bonding and structure in covalent molecules, as well as the mechanism of reactions.

Factors affecting lattice energy

1. The charge of the ions
2. The size of the ions

The higher the charge of the ions, the stronger the bond between them and the greater the lattice energy released.

The smaller the size of the ions is, the stronger the force of attraction between them, releasing more lattice energy.

Energy changes during the formation of aqueous solutions

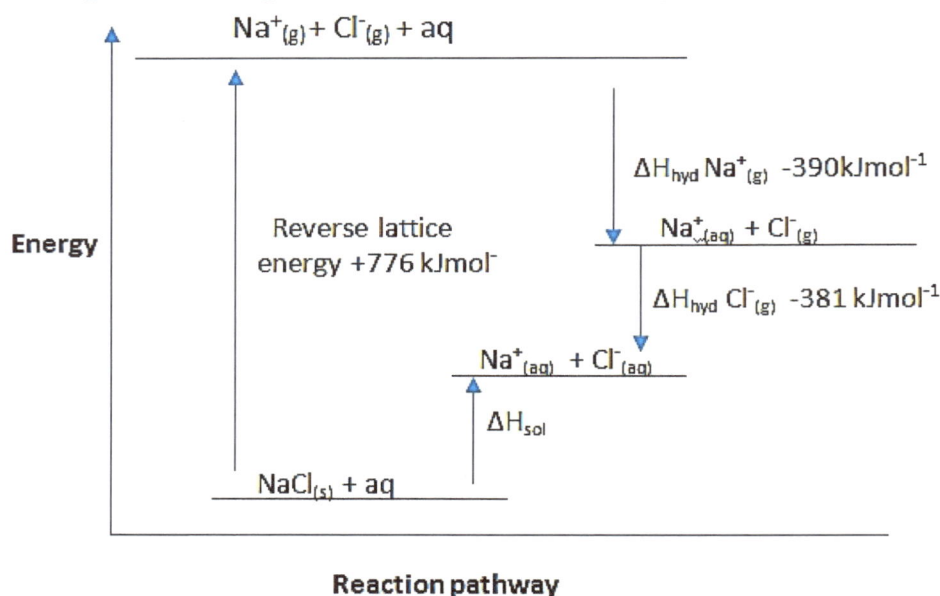

Reaction pathway

Fig. 9e energy cycle diagram for the formation of a solution of NaCl

Worked example 9.4

Calculate the enthalpy change of solution for sodium chloride.

ΔH_{sol} = Reverse lattice energy + $\Delta H_{hyd}Na^{+}_{(g)}$ + $\Delta H_{hyd}Cl^{-}_{(g)}$

= +776 + (-390) + (-381)

= **+5 kJmol^{-1}**

If a solid dissolves exothermically, it means that the substance is very soluble since it dissolves readily. The more negative the ΔH value is, the more soluble the solute is. However, the more energy that is absorbed for dissolving a solute, the less soluble it is, that is, the more positive the ΔH solution is, the less soluble the solute. If the hydration energy exceeds $\Delta H^{\ominus}_{latt}$, solubility occurs readily.

Trends in solubility among the Group 2 Sulphates

The solubility of the group 2 sulphates decreases from magnesium sulphate ($MgSO_4$) to barium sulphate ($BaSO_{4)}$. The **factors that affect the solubility** are:

1. **The charge on the ions**
2. **The size of the ions**

During hydration there is bonding between the ions and the H_2O molecules releasing **hydration energy**. If the hydration energy exceeds the reverse lattice energy that is required, solubility occurs readily.

The magnesium sulphate is most soluble in water and barium sulphate the least soluble. This is related to the **charge density** of the ions. Charge density is the ratio of the charge of an ion to its size. The Mg^{2+} ion is the smallest, with the highest charge density, and will therefore attract and bond most strongly with the water molecules, releasing the most hydration energy. This will more than compensate for the reverse lattice energy needed. Hence $MgSO_4$ dissolves readily and exothermically. $BaSO_4$ is least soluble because the Ba^{2+} ion is very large, with a small charge density. Therefore, it attracts and bonds to the water molecules most weakly, releasing small amounts of hydration energy which is unable to compensate for the reverse lattice energy. Hence ΔH_{soln} is very endothermic and barium sulphate is very sparingly soluble.

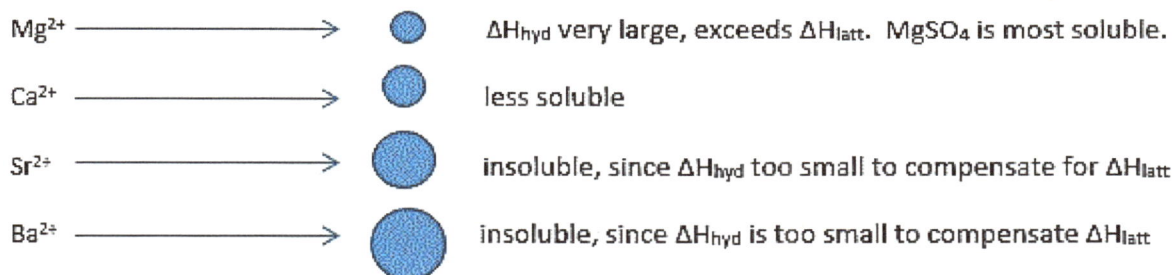

Fig. 9f *diagram relating the size of the cations to the solubility of Group 2 sulphates*

Energy Profile Diagrams

The energy profile diagram summarises all the energy absorbed and all the energy released during a chemical reaction.

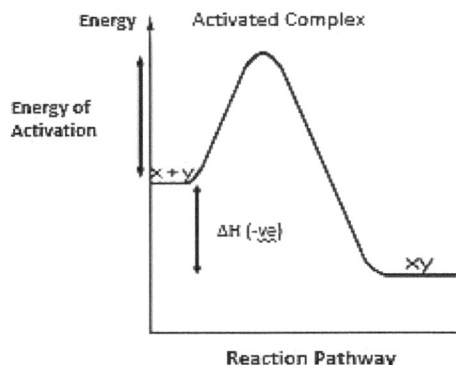

Fig. 9f *Energy profile diagram of an exothermic reaction*

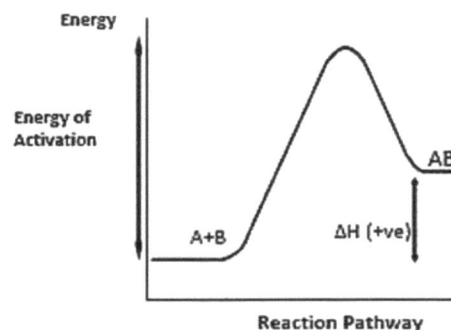

Fig. 9g *Energy profile diagram of an endothermic reaction*

The energy of activation is the minimum amount of energy required before the reaction can occur. It is the sum of all the energy absorbed during the reaction. The energy change of the reaction is the difference between all the energy absorbed and released.

Electron Affinity

The first electron affinity is always exothermic because an electron is attracted to the nucleus of a neutral atom. However, 2nd and subsequent electron affinities can be exothermic or endothermic because the negative electron is approaching a negative ion and they repel each other. Energy therefore has to be absorbed to force the two negative particles together. However, once they bond, energy is released. If the energy released is greater than the energy that was absorbed, then the second electron affinity would be exothermic. In contrast, if more energy was absorbed than was released, then overall the second electron affinity would be endothermic.

For example,

$O_{(g)} + e^- \rightarrow O^-_{(g)}$ $\Delta H_{e(1)}$ -ve

$O^-_{(g)} + e^- \rightarrow O^{2-}_{(g)}$ $\Delta H_{e(2)}$ +ve

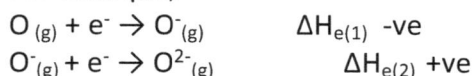

Experimental Lattice Energy vs. Theoretical Lattice Energy

	Experimental	Theoretical
NaCl	-776KJmol^{-1}	-766KJmol^{-1}
AgCl	-890KJmol^{-1}	-768KJmol^{-1}

This difference in experimental and theoretical lattice energy is due to an additional amount of energy released as a result of covalency in the ionic bond. If there was no covalent character in an ionic bond, then the experimental and theoretical lattice energies would be the same. The greater the difference, the greater the covalent character in the ionic bond. AgCl has more covalent character than NaCl.

Exercise 9.1

Calcium oxide, (quicklime) is produced by roasting limestone. Quicklime is used to neutralize the acidity of soils.

a. Define the term 'standard enthalpy of formation'. (2 marks)

b. Write a balanced equation for formation of $CaO_{(s)}$. (1 mark)

c. Construct a Born-Haber cycle for the formation of $CaO_{(s)}$ showing clearly, using equations, the steps of enthalpy changes involved. (3 marks)

d. Distinguish between 'exothermic enthalpy change' and 'endothermic enthalpy change', citing enthalpy changes from your Born-Haber cycle in (c) to support you answer. (2 marks)

e. Explain, by writing an equation, how the data from the Born-Haber cycle can be used to calculate the lattice energy of $CaO_{(s)}$. (2 marks)

9.2

The complete dissociation of a molecule of methane gas requires 1660kJ mol^{-1} of energy.

a. i) Define the term 'bond energy'. (2 marks)

ii) Calculate the bond energy of a C – H bond. (1 mark)

b. i) Explain why hydrogen chloride does not decompose on heating in the lab, whereas hydrogen iodide is decomposed by a hot wire. (2 marks)

_____

_____

ii) Determine with an explanation, which of the hydrogen halides, HBr or HI, would form $H^+_{(aq)}$ ions most readily in aqueous solution. (2 marks)

_____

_____

c. Describe what you would observe when the reaction in (b.i) above occurs, and write a balanced equation for the reaction. (2 marks)

_____

_____

d. i) Define the term 'standard enthalpy of combustion'. (2 marks)

_____

_____

ii) Write chemical equations to represent:
ΔH_f ($CO_{(g)}$) and ΔH_c ($CO_{(g)}$) (2 marks)

_____

_____

_____

e. i) Give the reason why the enthalpy change of formation of carbon monoxide cannot be determined directly. (2 marks)

_____

_____

9.3

a. Calculate the value for $\Delta H^{\varnothing}$ for the following reaction (3 marks)

$$CO_{(g)} + 1/2\ O_{2\ (g)} \rightarrow CO_{2(g)}$$

given: $\Delta H^{\varnothing}_f CO = -110.52$ kJ mol^{-1}
$\Delta H^{\varnothing}_f CO_2 = -393.51$ kJ mol^{-1}

b. Draw an energy cycle diagram to represent the reaction above. (2 marks)

Chapter 10

KINETIC THEORY OF GASES

Assumptions of the Kinetic Theory of Gases

1. Gases are made up of a large number of particles that are in continuous, random motion.
2. There are no forces of attraction between the particles of a gas.
3. The collisions of the particles are perfectly elastic; having no net gain or loss of energy.
4. The particles of a gas are so small and are so far apart that their contribution to the volume of a gas is negligible.

The above assumptions represent the behaviour of an 'ideal gas'.

The Gas Laws

Several scientists performed experiments with various gases to study the behaviour of gases.

Boyle's Law and Charles's Laws

These two scientists studied the effect of pressure and temperature respectively on the volume of a given mass of a gas.

Boyle's Law states that the volume of a given mass of gas is inversely proportional to the pressure at a constant temperature.

$$V \propto \frac{1}{P} \quad \text{hence } V = c\frac{1}{P} \quad \text{and } PV = c \quad \therefore P_1V_1 = P_2V_2$$

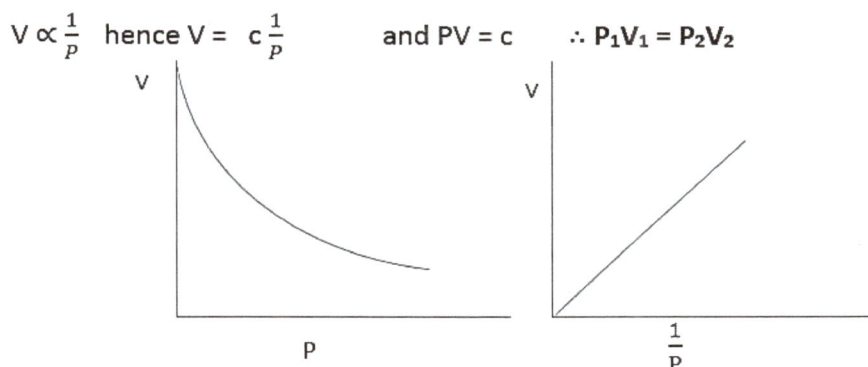

Fig.10a *graphs illustrating Boyle's Law*

Charles's Law states that the volume of a given mass of gas at a constant pressure is directly proportional to the absolute temperature (273K = 0°C).

$$V \propto T \quad \text{hence } V = cT \quad \text{and } \frac{V}{T} = C \quad \therefore \frac{V_1}{T_1} = \frac{V_2}{T_2}$$

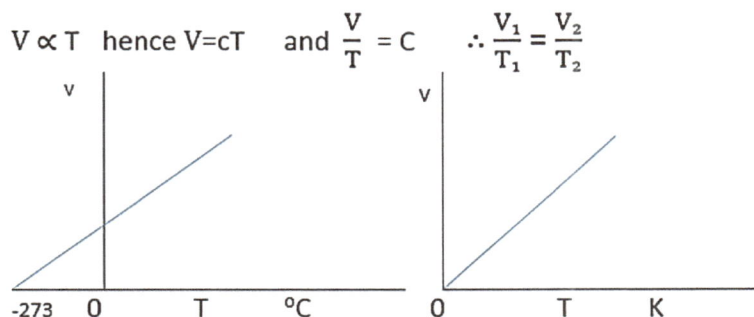

ig. 10b *graphs illustrating Charles's Law*

Charles and Boyle's Laws are combined to give the ideal gas equation:

75

PV = nRT

n = number of moles of gas

R = the gas constant

$$\frac{PV}{T} = R \qquad \frac{P_1V_1}{T_1} = \frac{P_2V_2}{T_2}$$

The equation PV = nRT is most widely used to calculate the relative molecular masses of gases and volatile liquids.

$$PV = \frac{mRT}{M} \qquad M = \frac{mRT}{PV}$$

Application of the kinetic theory to the gas laws

Charles' Law

Temperature changes will impact on the rate of collision of the particles and the spread of the particles. By increasing the temperature, the kinetic energy of the particles increases and they move farther apart, increasing the volume. This behaviour is related to the assumptions that the particles have no forces of attraction between them, as well as the particles contribute negligibly to the total volume of the gas.

Boyle's Law

The particles of the ideal gas are far apart with no forces of attraction between them. The gas is therefore compressible. At high pressure the particles are forced closer together, occupying a smaller volume, as shown in the diagrams below. When the pressure is doubled, the volume of the gas is halved.

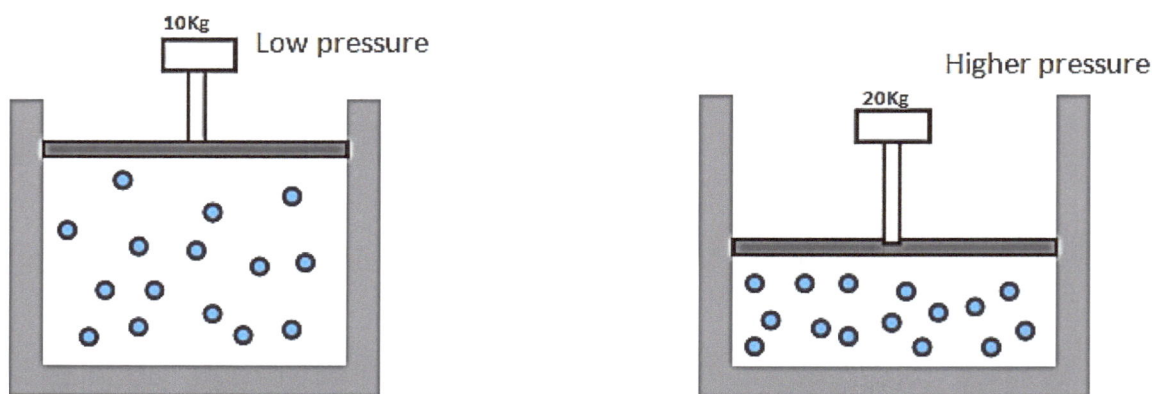

Fig. 10c *showing the effect of pressure on the volume of a gas*

Deviation of 'Real Gases' from 'Ideal Gas' Behaviour

An ideal gas obeys the gas laws and the assumptions of the Kinetic Theory.

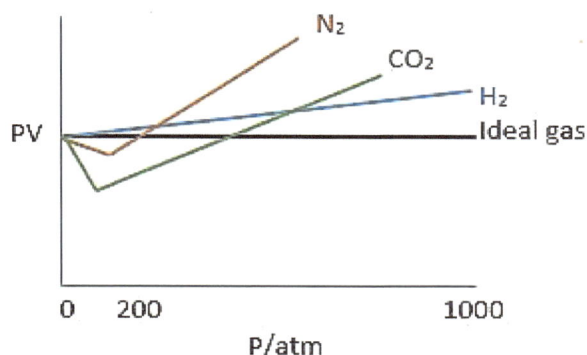

Fig.10d *deviation of real gases from ideal behaviour*

The plot of PV against P shows that the gases deviate from ideal behaviour.

As gases deviate from ideal gas behaviour, they are referred to as **'real gases'**. This deviation occurs at **high pressure** and **low temperature.** At high pressures the particles of the gas are forced so close together that intermolecular forces of attraction develop between them, resulting in the gas liquefying.

Similarly, at low temperatures the particles lose kinetic energy and as they come to lie close together, intermolecular forces of attraction develop between them, and the gas liquifies. A combination of high pressure and low temperature causes liquefaction of gases. As the temperature of a gas is increased, the amount of pressure needed to bring about liquefaction has to be increased. There is a temperature that is so high and the particles are so far apart that no amount of increased pressure can cause liquefaction. This temperature is called the **'critical temperature'**. The minimum pressure required to liquefy a gas at its critical temperature is called its **'critical pressure'**.

P	atm	kPa/kNm^{-2}	Pa/Nm^{-2}
V	dm^3	dm^3	m^3
T	K	K	K
R	0.0821	8.31	8.31
m	g	g	g

Table 10.1 *Combination of units for calculations using the ideal gas equation*

Note. 1 atm = 760mm Hg =101.325 kPa = 101325 Pa

Worked example 10.1

0.275g of a volatile liquid 'A' has a volume of 75cm^3 at 127 °C and 1 atm. Calculate the relative molecular mass of 'A'.

Solution

$$M = \frac{mRT}{PV}$$

$$= \frac{0.275 \times 0.0821 \times (127 + 273)}{1 \times \left(\frac{75}{1000}\right)} = 120.4$$

The liquid state

The particles of a liquid have bonds between them. Unlike solids, however, the particles are not arranged in any definite manner, and they possess sufficient kinetic energy to allow them to move freely around each other. The liquid therefore flows easily, resulting in the property known as **fluidity.**

Freezing is a change of state from liquid to solid at decreased temperatures. **Melting** is a change of state from solid to liquid on increasing the temperature.

The temperature at which the solid and liquid states of an element or compound, are in equilibrium with each other at atmospheric pressure, is known as its **melting / freezing point.** The melting / freezing point of water is zero/ 0°C.

Some solids on heating sublime at atmospheric pressure and therefore melt at higher pressures. Examples are iodine, diamond and graphite.

The change of state from liquid to vapour at increased temperatures is known as **vaporisation.** The bonds between the particles of the liquid are broken and the evaporation of free gaseous particles occurs. **Volatility** of a liquid is a measure of its tendency to vaporise. Very volatile liquids have weak bonds between the particles. For example, liquids with intermolecular forces of attraction such as tetrachloromethane, gasoline and water are volatile.

Non-volatile liquids have strong bonds between the particles, for example mercury, which has strong metallic bonds between its atoms.

The temperature at which a liquid and its vapour are in equilibrium with each other, at atmospheric pressure is known as the **boiling point.** The boiling point of water is 100°C.

Simple Laboratory Determination of boiling Point and Melting point

Boiling Point determination

Place about 10 cm^3 of the liquid under investigation, into a boiling tube and heat carefully over a low Bunsen flame or in a boiling water bath. A boiling stone may be placed in the tube to minimise 'bumping' of the boiling liquid.

To determine the boiling point of a pure substance, hold the thermometer bulb above the liquid and record the constant temperature of the vapour.

To determine the boiling point of an impure liquid, insert the thermometer bulb into the mixture and record the highest constant temperature.

Freezing/Melting Point Determination

Pack a melting point **glass capillary tube** with a small amount of the dry solid under investigation. Using a rubber band, attach it to a thermometer so that solid sample is lined up with the middle of the thermometer bulb. Place this in an oil bath and heat gently.

Record the temperature when the first drop of liquid is observed and a second temperature when the entire solid has melted. Record a melting range. Pure substances usually have a sharp melting point and a narrow range, for example 120°C – 120.5°C while impure samples have a wider temperature range, for example 135°C – 138°C.

An alternative method is to place about 1 g of the dry solid sample in **a hard glass test tube**. Heat until the solid melts. Remove the tube from the heat source, insert a thermometer bulb in the liquid, stir gently and record the constant temperature over which the liquid solidifies.

Exercise 10.1
a. State three assumptions of the kinetic theory of an 'ideal gas'. (2 marks)

_____

_____

b. Distinguish between an 'ideal gas' and a 'real gas' and state two assumptions of the kinetic theory that do not hold true for a 'real' gas. (2 marks)

_____

_____

_____

c. A gas occupies 140 cm³ at a pressure of 0.88 atm. What is the new volume of the gas if it is allowed to expand at a pressure of 0.56 atm at a constant temperature of 25°C? (2 marks)

10. 2.

a. State Charles's Law and use it to explain how a hot air balloon works. (4 marks)

_____

_____

_____

_____

_____

b. What is the volume of a gas in a balloon at 195°C if the balloon was filled to a volume of 5dm³ at 25°C? (2 marks)

c) Use the 'ideal gas' equation to calculate the number of moles of air molecules the hot air balloon contains at 195°C and a pressure of 1.01 x 10^5 Pa. (2 marks)

Chapter 11

Chemical Equilibria

The word equilibrium means *balance*. Equilibrium can be static or dynamic. There is **movement** occurring during **dynamic equilibrium**. An example is when the rate of evaporation of a liquid, equals the rate of condensation of its vapour in a closed container. Both processes are occurring, but at equal rates.

There is **no movement** during **static equilibrium**. The sum of opposing forces, about an equilibrium point is zero. For example, when children of equal masses are on opposite ends of a 'see-saw', the masses are balanced and there is zero motion.

Dynamic equilibrium is exhibited in many chemical reactions. The topic of dynamic equilibrium relates to **reversible reactions**. In a reversible reaction, reactants form products and the products form reactants in the same reaction system.

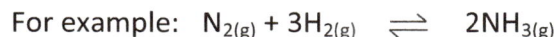

$$\text{For example:} \quad N_{2(g)} + 3H_{2(g)} \rightleftharpoons 2NH_{3(g)}$$

Initially, there is 100% of reactants and so the rate of the forward reaction (left to right) is very fast, producing a small amount of product, which at a slower rate, produces reactants in the backward reaction. As time progresses, the concentration of the reactants decreases while the concentration of the product increases. Therefore, the rate of the forward reaction is decreasing, whilst the rate of the backward reaction is increasing. Eventually, the rates of the forward and backward reactions become equal. When this occurs, the system is said to be in **dynamic equilibrium**.

A system in dynamic equilibrium has the following characteristics:
- The concentration of the reactants and products remains constant.
- Macroscopic processes remain constant (visible changes).
- Microscopic changes are still occurring at the molecular level.
- The equilibrium can only be achieved in a closed container/vessel.
- The equilibrium can be obtained from either direction of the reaction system.

For any system in dynamic equilibrium, **the ratio of the product of the concentration of the products raised to the power of their coefficient to the product of the concentration of the reactants, raised to the power of their coefficient in the balanced equation, is constant.** This constant is called the **equilibrium constant** and it is designated K. The above statement is called the **Equilibrium Law.**

Example
The Equilibrium constant expression for the reaction $\quad N_{2(g)} + 3H_{2(g)} \rightleftharpoons 2NH_{3(g)}$
is:

$$\frac{[NH_3]^2}{[N_2][H_2]^3} = K_c$$

Note [ ] *means concentration of, in moldm^{-3}.*
For gases, the concentration can also be expressed in terms of the partial pressure of the gas. The concentration/[ ] of a gas is proportional to the partial pressure. The **partial pressure** of a

gas is the pressure that an individual gas exerts in a mixture of gases. Therefore, the higher the concentration, the higher the partial pressure of the gas.

The equilibrium constant expression for the above ammonia system can be expressed in partial pressures as follows: $K_p = \dfrac{(pNH_3)^2}{(pN_2)(pH_2)^3}$ (p= partial pressure)

In the ammonia system, the reactants and products are in the same state of matter. This is an example of a **homogenous equilibrium system.**

However when different states exist among the reactants and products, the system is described as **heterogenous equilibrium.**

For example:

$CaCO_{3(s)} \rightleftharpoons CaO_{(s)} + CO_{2(g)}$

$[CO_2] = K_c$

$(pCO_2) = K_p$

Note: *pure solids and liquids have a constant concentration, so solid CaO and solid CaCO₃ have a constant concentration assigned 1.*

Worked example 11.1

An equilibrium mixture contains 2 moles of Br_2, 1.25mol of H_2 and 0.5mol HB_r in a $4dm^3$ vessel. Calculate the value of K_c for the reaction: $H_{2(g)} + Br_{2(l)} \rightleftharpoons 2HBr_{(g)}$

Solution

$$K_c = \dfrac{[HB_r]^2}{[H_2][Br_2]}$$

$[HBr] = 0.5mol \div 4dm^3 = 0.125$ moldm^{-3}

$[Br_2] = 2mol \div 4dm^3 = 0.5$ moldm^{-3}

$[H_2] = 1.25mol \div 4dm^3 = 0.3125$ moldm^{-3}

$K_c = \dfrac{(0.125)^2}{0.3125 \times 0.5}$ **calculation of units:** $\dfrac{moldm^{-3} \times moldm^{-3}}{moldm^{-3} \times moldm^{-3}}$ These unit cancel out

Answer = **0.1 (no units)**

Worked example 11.2

For the following reaction at equilibrium: α d glucose $\rightleftharpoons$ β d glucose, K_c is 1.75. If the initial α d glucose concentration is 0.01 moldm^{-3}, calculate the β d glucose concentration at the same temperature at which K_c was determined.

Solution

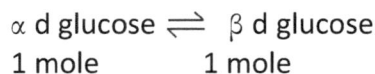
α d glucose $\rightleftharpoons$ β d glucose

1 mole   1 mole

[Initial /moldm^{-3}] 0.01   0

If y mole of α d glucose reacts,

At equilibrium 0.01 – 1y α d glucose + 1y β d glucose are in the equilibrium mixture

$K_c = \dfrac{[\beta\, d\, glucose]}{[\alpha\, d\, glucose]}$

$1.75 = \dfrac{1\,y}{0.01 - y}$

$0.0175 - 1.75\,y = 1\,y$

$0.0175 = 2.75\,y$

Answer   y = **0.006moldm^{-3}**

Worked example 11.3

When a 0.0500 mol sample of SO_3 is introduced into a 2dm^{-3} vessel at 900 K, 0.015mol SO_3 is found to be present at equilibrium. Determine K_c for the decomposition of SO_3 at 900 K using the following equation:

Solution

$$2SO_{3(g)} \rightleftharpoons 2SO_{2(g)} + O_{2(g)}$$

mole ratios   2 moles   2 moles   1 mole

[initial]/moldm^{-3} 0.05/2= 0.025   0   0

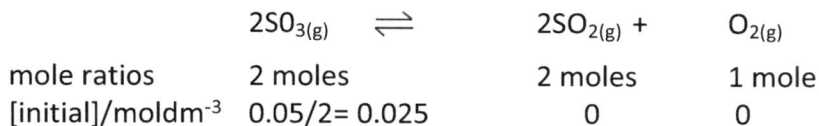

At equilibrium

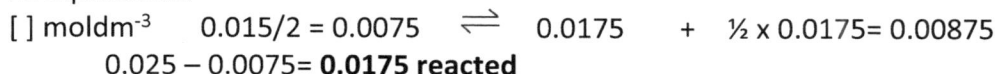

[ ] moldm^{-3} 0.015/2 = 0.0075 $\rightleftharpoons$ 0.0175   +   ½ x 0.0175= 0.00875

        0.025 − 0.0075= **0.0175 reacted**

$$K_c = \frac{[SO_2]^2[O_2]}{[SO_3]^2}$$

$$K_c = \frac{(0.0175)^2 \times (0.00875)}{(0.0075)^2}$$

Answer K_c = **0.048 moldm^{-3}**

Worked example 11.4

N_2O_4 is 30% dissociated at 101.325kPa. Calculate Kp.

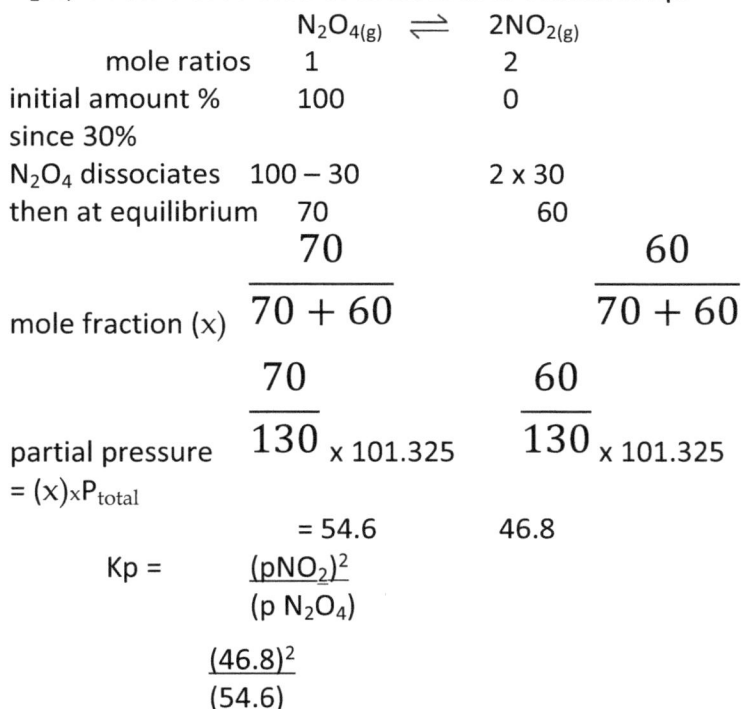

$$N_2O_{4(g)} \rightleftharpoons 2NO_{2(g)}$$

      mole ratios   1   2

initial amount %   100   0

since 30%

N_2O_4 dissociates   100 − 30   2 x 30

then at equilibrium   70   60

mole fraction (x) $\dfrac{70}{70 + 60}$ $\dfrac{60}{70 + 60}$

partial pressure $\dfrac{70}{130}$ x 101.325 $\dfrac{60}{130}$ x 101.325

= (x)$_x$P$_{total}$

           = 54.6   46.8

    Kp = $\dfrac{(pNO_2)^2}{(p\,N_2O_4)}$

       $\dfrac{(46.8)^2}{(54.6)}$

Answer = **40.1 kPa**

Exercise 11.1

Calculate the pressure at which N_2O_4 will be 40% dissociated at the same temperature.

Le Chatelier's Principle and factors affecting equilibria

Le Chatelier's Principle states that if a system in equilibrium is subject to a change, the system will react in such a way as to counteract or nullify the change and establish a new equilibrium.

Factors that will affect a system in equilibrium are:
1. Concentration of reactants and products
2. Pressure for a gaseous system
3. Presence of a catalyst
4. Temperature

In considering how these factors affect an equilibrium system, we must look at:
1. How the equilibrium position is affected, that is whether it causes the equilibrium to shift to the left or to the right.
2. How the equilibrium composition is affected.
3. How the equilibrium constant is affected.

Concentration

Consider the system below:

$$N_{2(g)} + 3H_{2(g)} \rightleftharpoons 2NH_{3(g)}$$

If the concentration of ammonia is decreased, according to Le Chatelier's Principle, the system will react to increase it. Hence, the equilibrium position shifts to the right. When a new equilibrium is established, the system will contain less of the reactants nitrogen and hydrogen and more of the product, ammonia. However, the proportion of the reactants and products remains the same. Therefore, K_c remains unchanged.

If the concentration of nitrogen or hydrogen is decreased, the equilibrium position will shift to the left. When a new equilibrium is established, the concentration of ammonia will be decreased. K_c remains constant since the proportion of the reactants and product remains unchanged.

Pressure

Consider the ammonia system above. If the pressure on the system is increased, the system will react to decrease the pressure. Hence, the equilibrium position shifts to the right where there is a lower partial pressure due to 2 moles of gaseous products, as compared to 4 moles of gaseous reactants on the left. When a new equilibrium is established, the concentration of nitrogen and hydrogen would have decreased, whilst the concentration of ammonia would have increased. However, K_c and K_p remains the same.

When the pressure on the system is decreased, the equilibrium position shifts to the left where there is a higher number of moles of gas. When a new equilibrium is established, there would be lower concentrations of NH_3 and higher concentrations of H_2 and N_2. K_c remains the same.

Catalysts

Catalysts alter the rates of the chemical reactions. Positive catalysts increase the rate and negative catalysts decrease the rate. A positive catalyst speeds up the rate of the forward and backward reactions to the same extent. Hence, there would be no effect on the equilibrium position. It doesn't change the equilibrium constant or the equilibrium composition. It simply causes the equilibrium to be established faster.

Temperature

Consider the system below:

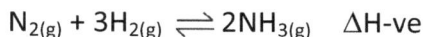

$$N_{2(g)} + 3H_{2(g)} \rightleftharpoons 2NH_{3(g)} \quad \Delta H\text{-ve}$$

The forward reaction is exothermic and therefore the backward reaction is endothermic. If the temperature is increased, the rate of the backward endothermic reaction would be significantly increased over the rate of the forward exothermic reaction. Increased temperature favours the endothermic reaction but not the exothermic reaction. Hence, when a new equilibrium is established, there would be a very high $[N_2]$ and high $[H_2]$ and a low $[NH_3]$ in the equilibrium mixture. K_c and Kp would therefore decrease.

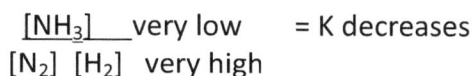

$$\frac{[NH_3] \quad \text{very low}}{[N_2]\ [H_2] \quad \text{very high}} = \text{K decreases}$$

On decreasing the temperature, the rate of the forward exothermic reaction far exceeds the rate of the reverse (backward) endothermic reaction. Hence, when a new equilibrium is established, there would be a very high $[NH_3]$ in the equilibrium mixture and a very low $[N_2]$ and $[H_2]$ so the value of K increases.

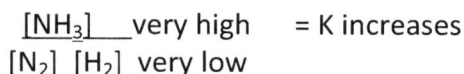

$$\frac{[NH_3] \quad \text{very high}}{[N_2]\ [H_2] \quad \text{very low}} = \text{K increases}$$

In the **Haber process,** for the industrial production of ammonia, low temperature and high pressure theoretically produce a high yield of NH_3. However, at low temperature, the rate of the reaction is too slow to be economical. High pressure is expensive to generate because of the cost of energy needed to generate it, as well as the cost of strong, reinforced, expensive equipment needed to withstand the high pressure. There is also a risk of explosion in the workplace, which is dangerous to the workers.

The compromising conditions that give the most economical yield of ammonia are 450°C and an average pressure of about 250 atm using finely divided iron as a catalyst along with 2 promoters of the catalyst K_2O and Al_2O_3. Under the above conditions, the maximum percentage of ammonia in the equilibrium mixture is 13 – 15%.

Ionic Equilibria

In ionic equilibria, reversible reactions containing ions are being studied.
These include:
- Weak acids
- Weak bases
- Water
- Saturated solutions

Weak acid/Weak base Equilibria

Bronsted and Lowry define **an acid** as a proton donor and **a base** as a proton acceptor.

Weak acid

Consider the following weak acid equilibrium:

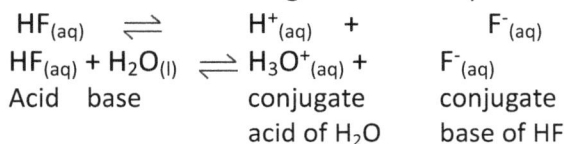

$HF_{(aq)} \rightleftharpoons H^+_{(aq)} + F^-_{(aq)}$

$HF_{(aq)} + H_2O_{(l)} \rightleftharpoons H_3O^+_{(aq)} + F^-_{(aq)}$

Acid   base   conjugate   conjugate

                   acid of H_2O base of HF

HF donates a proton to H_2O. Therefore HF is an acid and H_2O is a base in the forward reaction. In the reverse reaction, H_3O^+ donates a proton to F^-. Therefore H_3O^+ is referred to as the conjugate acid of H_2O whilst the F^- is the conjugate base of HF as it accepts a proton in the reverse reaction.

At equilibrium, the ratio of the product of the concentration of the hydroxonium ion (H_3O^+) and the conjugate base F^-, to the concentration of the unionised acid HF, all raised to the power of their coefficient in the ionic equation is constant. This constant is called the **acid dissociation constant, K_a**. It is constant at a constant temperature.

$$\frac{[H_3O^+_{(aq)}][F^-_{(aq)}]}{[HF_{(aq)}]} = K_a$$

OR

$$\frac{[H^+_{(aq)}][F^-_{(aq)}]}{[HF_{(aq)}]} = K_a$$

Acid	K_a at 25°C
HCl	large
HF	6.6×10^{-4}
CH_3COOH	1.8×10^{-5}
H_2CO_3	4.5×10^{-7}
HCO_3^-	4.7×10^{-11}

The K_a value is an indicator of the strength of an acid and can be used to compare the strengths of acids. The higher the K_a value is, the stronger the acid.

Note

$$H_2CO_{3(aq)} \rightleftharpoons H^+_{(aq)} + HCO_3^-{}_{(aq)} \quad K_a = 4.5 \times 10^{-7}$$
acid \qquad\qquad\qquad conjugate base

$$HCO_3^-{}_{(aq)} \rightleftharpoons H^+_{(aq)} + CO_3^{2-}{}_{(aq)} \qquad K_a = 4.7 \times 10^{-11}$$
acid \qquad\qquad\qquad conjugate base

The stronger an acid, the weaker its conjugate base and vice versa.

Worked example 11.5
Calculate the concentration of the $H^+_{(aq)}$ ion in a 0.1moldm⁻³ solution of ethanoic acid.

Solution

$$CH_3COOH_{(aq)} \rightleftharpoons CH_3COO^-{}_{(aq)} + H^+_{(aq)}$$

[initial/moldm⁻³] \qquad 0.01 \qquad\qquad 0 \qquad\qquad 0

If x moldm⁻³ of acid ionize, then at equilibrium

\qquad\qquad\qquad 0.1 − x \qquad\qquad x \qquad\qquad x

$$K_a = \frac{[CH_3COO^-][H^+_{(aq)}]}{[CH_3COOH_{(aq)}]}$$

$$1.8 \times 10^{-5} = \frac{x^2}{0.1 - x} \qquad \text{(Ignore − x since } K_a \text{ is very small)}$$

$$1.8 \times 10^{-5} = \frac{x^2}{0.1}$$

Answer $x = [H^+_{(aq)}] = 1.34 \times 10^{-3}$ **moldm⁻³**

Note *if K_a is $x~10^{-3}$ or less then (+/− x) is mathematically negligible and can be ignored.*

Weak base

Consider the weak base ammonia.

$$NH_{3(aq)} + H_2O_{(l)} \rightleftharpoons NH_4^+{}_{(aq)} + OH^-{}_{(aq)}$$
base \qquad acid \qquad\qquad conjugate \qquad conjugate base
\qquad\qquad\qquad\qquad\qquad acid of NH_3 \qquad of H_2O

Ammonia (NH_3) accepts a proton from water (H_2O) in the forward reaction. Therefore, ammonia is the base and water is the acid. In the reverse reaction NH_4^+ donates a proton to OH^-. Therefore, NH_4^+ is the conjugate acid of NH_3 and OH^- the conjugate base of water.

At equilibrium, the ratio of the product of the concentration of the $OH^-_{(aq)}$ ions and the concentration of the conjugate acid NH_4^+, to the concentration of the unionized base NH_3, all raised to the power of their coefficients in the ionic equation is constant. This constant is called the **base dissociation constant, K_b.**

$$K_b = \frac{[NH_4^+_{(aq)}][OH^-_{(aq)}]}{[NH_{3(aq)}]} \qquad K_b = 1.8 \times 10^{-5}$$

K_b, like K_a, indicates the strength of the base.

HCO_3^- is stronger as a base than as an acid, because its K_b value of 2.6×10^{-8} is higher than its K_a value of 4.7×10^{-11}.

Base	K_b at 25°C
O^{2-}	1×10^{22}
OH^-	55
NH_3	1.8×10^{-5}
HCO_3^-	2.6×10^{-8}
F^-	1.5×10^{-11}

Like acids, a strong base produces a very weak conjugate acid.

The Auto Ionization of Water

The purest H_2O will record some conduction of electricity, which means that ions are present. In water, one molecule behaves like an acid, donating a proton whilst another molecule behaves like a base, accepting a proton.

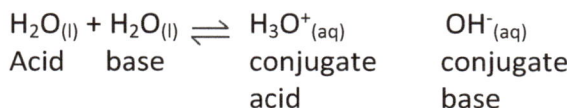

$$H_2O_{(l)} + H_2O_{(l)} \rightleftharpoons H_3O^+_{(aq)} \qquad OH^-_{(aq)}$$
Acid   base   conjugate   conjugate
                      acid   base

OR

$$H_2O_{(aq)} \rightleftharpoons H^+_{(aq)} + OH^-_{(aq)}$$

At equilibrium the product of the concentration of the aqueous hydrogen ion ($H^+_{(aq)}$) and the aqueous hydroxide ion ($OH^-_{(aq)}$) is constant. This constant called the **ionic product constant of water,** K_w which is equal to 1×10^{-14} at 25°C.

$$K_w = [H^+_{aq}][OH^-_{aq}]$$

In a solution, if the $[H^+_{aq}] = [OH^-_{aq}]$, the solution is said to be **neutral.**
However if $[H^+_{aq}] > [OH^-_{aq}]$, the solution is **acidic.** If $[H^+_{aq}] < [OH^-_{aq}]$, the solution is **basic.**

Worked example 11.6
Calculate the values for the aqueous hydrogen ion and hydroxide ion concentration, in a neutral solution at 25°C.

Solution   Since $[H^+_{aq}] = [OH^-_{aq}]$ in a neutral solution
                      Then $[H^+_{aq}][OH^-_{aq}] = K_w = 1.0 \times 10^{-14}$
                      Let a be [ions]
                      $a \times a = 1.0 \times 10^{-14}$
                      $a = \sqrt{1.0 \times 10^{-14}}$

                    Answer $= 1.0 \times 10^{-7}$ moldm^{-3}

Worked example 11.7
By considering the ions produced in aqueous solution, determine if the following solutions will be acid, basic or neutral: $NaCl$, NH_4Cl and $NaHCO_3$.

Solution

	Na$^+$	Cl$^-$	Therefore
NaCl	(very weak acid) Conjugate acid of strong base NaOH	(very weak base) Conjugate base of strong acid HCl	NaCl is neutral
NH$_4$Cl	NH$_4^+$ (stronger as an acid) Conjugate acid of weak base NH$_3$	Cl$^-$ (very weak base) Conjugate base of strong acid HCl	Therefore NH$_4$Cl is acidic
NaHCO$_3$	Na$^+$ (very weak acid) Conjugate acid of strong base NaOH	HCO$_3^-$ (stronger as a base) Conjugate base of weak acid H$_2$CO$_3$	Therefore NaHCO$_3$ is basic.

The Relationship Between K$_a$, K$_b$ and K$_w$

Consider HF$_{(aq)}$ and its conjugate base F$^-$ in aqueous solutions:

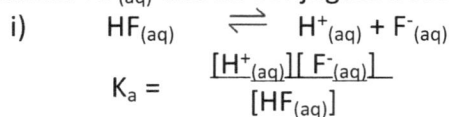

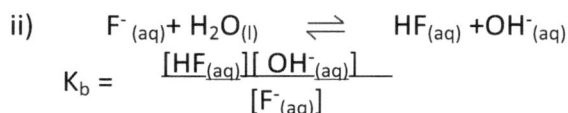

i) $HF_{(aq)} \rightleftharpoons H^+_{(aq)} + F^-_{(aq)}$

$$K_a = \frac{[H^+_{(aq)}][F^-_{(aq)}]}{[HF_{(aq)}]}$$

ii) $F^-_{(aq)} + H_2O_{(l)} \rightleftharpoons HF_{(aq)} + OH^-_{(aq)}$

$$K_b = \frac{[HF_{(aq)}][OH^-_{(aq)}]}{[F^-_{(aq)}]}$$

∴   By multiplying (i) and (ii), $K_a \times K_b = [H^+_{(aq)}][OH^-_{(aq)}] = K_w = $ **1.0 x 10^{-14}**

Worked example 11.8

K$_a$ for HF is 6.8 x 10^{-4}. Find K$_b$ for its conjugate base F$^-$.

Solution

$K_w = K_a \times K_b$

$\dfrac{K_w}{K_a} = K_b$ ∴ $\dfrac{1.0 \times 10^{-14}}{6.8 \times 10^{-4}} = $ **1.47 x 10^{-11}**

pX Scales

'p' in pX means "$-\log_{10}$ of".

X is H in pH, OH in pOH, K$_a$ in pK$_a$, K$_b$ in pK$_b$ and K$_w$ in pK$_w$.
Therefore:
pH is defined as the negative $\log_{10}$ of the aqueous H$^+$ ion concentration of a solution, mathematically expressed as pH = $-\log_{10}$ of [H$^+_{aq}$]
pOH is the negative $\log_{10}$ of the aqueous OH$^-$ concentration of a solution: = $-\log_{10}$ of [OH$^-_{aq}$]
pK$_a$ is the negative $\log_{10}$ of the acid dissociation constant: = $-\log_{10}$ K$_a$
pK$_b$ is the negative $\log_{10}$ of the base dissociation constant: = $-\log_{10}$ K$_b$
pK$_w$ is the negative $\log_{10}$ of the ion product constant of water: = $-\log_{10}$ K$_w$

pH

Worked example 11.9

Calculate the pH of a neutral solution at 25°C.

Solution

$$pH = -\log_{10} [H^+_{(aq)}]$$

in a neutral solution: $[H^+_{aq}] = [OH^-_{aq}]$

$$[H^+_{aq}] \times [OH^-_{aq}] = 1.0 \times 10^{-14}$$

$$\therefore [H^+_{aq}] = \sqrt{1.0 \times 10}^{-14}$$
$$= 1.0 \times 10^{-7}$$
$$pH = -\log_{10} (1.0 \times 10^{-7})$$

Answer = **7**

Worked example 11.10

Calculate the pH of a solution containing 0.01moldm^{-3} of OH$^-_{(aq)}$ ions.

Solution

$$[H^+_{aq}] \times [OH^-_{aq}] = 1.0 \times 10^{-14}$$
$$[H^+_{aq}] \times 0.01 = 1.0 \times 10^{-14}$$
$$[H^+_{aq}] = \frac{1.0 \times 10^{-14}}{0.01}$$
$$= 1.0 \times 10^{-12}$$
$$pH = -\log_{10} [H^+_{aq}]$$
$$= -\log_{10} (1.0 \times 10^{-12}) \qquad Answer = \mathbf{12}$$

Note *that pH + pOH = pK$_w$ = 14*

An alternative method for example 11.10 is as follows:

$$pOH = -\log_{10} 0.01 = 2$$
$$\therefore pH = 14 - 2$$

Answer = **12**

Worked example 11.11

A sample of fresh apple juice has a pH of 3.76. Calculate the H$^+_{(aq)}$ concentration of the juice

Solution

$$pH = -\log_{10} [H^+_{aq}]$$
$$[H^+_{aq}] = 10^{-pH}$$
$$= 10^{-3.76} \qquad \textbf{\textit{Answer}} [H^+_{aq}]= \mathbf{1.74 \times 10^{-4}\ moldm^{-3}}$$

Measuring pH

The pH of a solution can be accurately measured using a **pH meter**. The instrument gives rapid and accurate pH readings. A pH meter consists of a pair of electrodes which are placed in the solution to be measured. The electrodes form an electrochemical cell with the solution and a voltage is generated, which is proportional to the [H$^+_{aq}$] of the solution.

Methodology

First the pH meter is calibrated by placing the electrodes in solutions of known [H$^+_{aq}$] and recording the pH. Once it is calibrated, the scale on the pH meter will read off pH values. The electrodes are then withdrawn, washed with distilled water and placed in the test solution. The pH for the test solution will be read off.

Titration curves and choice of indicators used in acid-base titration

During a titration, a solution in a burette, usually the standard solution called the titrant is added drop by drop to another accurately measured solution, the tirand in a conical flask. In an acid-base titration, $H^+_{(aq)}$ ions and $OH^-_{(aq)}$ ions react and form neutral water in a neutralization reaction.

The volume of an acid/base that will exactly neutralize the base/acid in the conical flask, in accordance to the mole ratio in a balanced equation between the acid and base, is called the **equivalence point**.

During an acid-base titration, an indicator is added to the solution in the conical flask to detect the end-point of the reaction. An **indicator** is a compound which is one colour in an acid and another colour in a base. The equivalence point is undetectable and so a small excess of acid/base is added to the base/acid in the conical flask. A colour change is observed. The total volume that brings about the colour change is called the **endpoint** of the titration. At the end-point of a titration there is a physical change. This could be a change in the colour of an indicator; a change in temperature, a change in pH, or a change in turbidity.

Common laboratory indicators are methyl orange and phenolphthalein. Methyl orange is yellow in basic conditions and red in acid conditions. Phenolphthalein is colourless in acid and red in basic conditions. Indicators have a range of pH values over which they change colour and this is called the **pH range**.

The pH range of methyl orange is 3 – 5. The pH range of phenolphthalein is 8 – 10.

Titration Curves

These show the pH changes that occur during the course of an acid - base titration. From these curves, the best indicator can be chosen based on the pH range.

Below are **titration curves for a monobasic acid and a monoacidic base of different strengths.**

For example, $HCl_{(aq)} + NaOH_{(aq)} \rightarrow NaCl_{(aq)} + H_2O_{(l)}$ (a strong acid and a strong base).

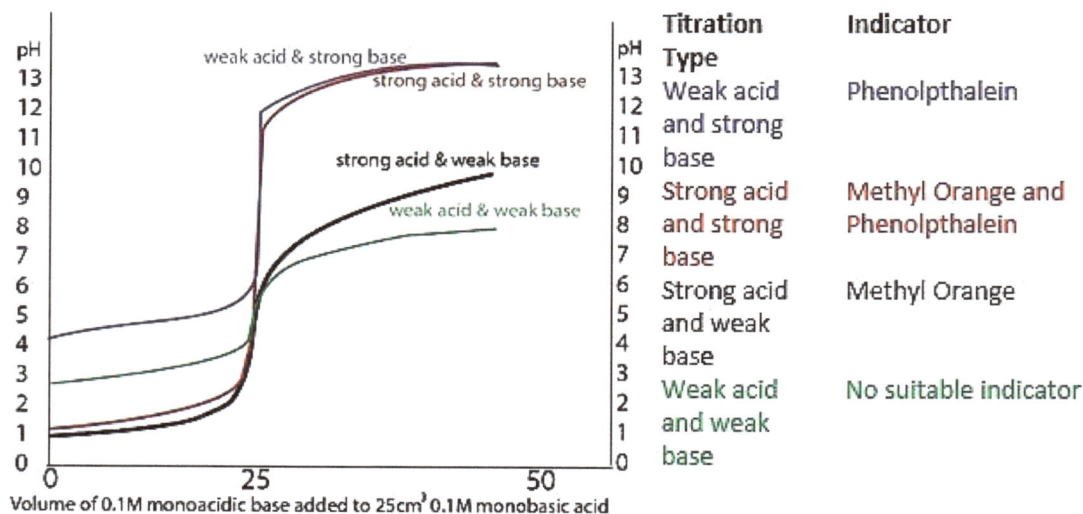

Volume of 0.1M monoacidic base added to 25cm³ 0.1M monobasic acid

Titration Type	Indicator
Weak acid and strong base	Phenolpthalein
Strong acid and strong base	Methyl Orange and Phenolpthalein
Strong acid and weak base	Methyl Orange
Weak acid and weak base	No suitable indicator

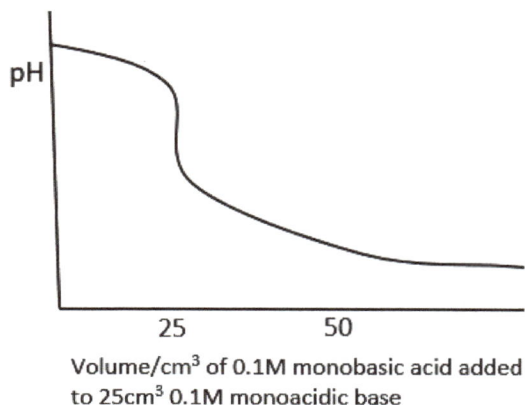

Fig. 11b *Titrating acid against base*

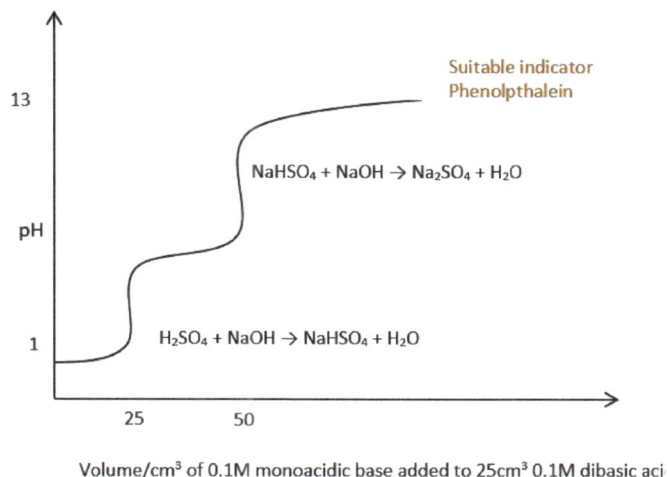

Fig.11c *Titrating a monoacidic base against a dibasic acid*

e.g. $H_2SO_4 + 2NaOH \rightarrow Na_2SO_4 + 2H_2O$

pK_a and pK_b **values** are often used to compare the strengths of acids and bases respectively. The stronger an acid or base is, the higher the K_a/K_b values and the lower the pK_a/pK_b values.

Buffer Solutions

A buffer solution is defined as one that resists change in pH on adding small amounts of an acid or a base. A buffer solution is made from a weak acid and its conjugate base or a weak base and its conjugate acid.

Consider the solutions **a, b** and **c** respectively, below. Only b is a buffer solution. On adding 1 cm^3 of dilute HCl, to each solution, the pH of solution b decreases only very slightly. Similarly, on adding $1cm^3$ of $NaOH_{(aq)}$, the pH of solution b increases only very slightly as compared to the large pH changes in solutions a and c.

	solution a	solution b	solution c
	CH_3COOH	$CH_3COO^-\ H^+$ $CH_3COO^-\ Na^+$	$CH_3COO^-\ Na^+$
Initial pH	4.20	6.50	7.60
pH on adding 1 cm^3 HCl acid	1.90	6.49	2.90
pH on adding $1cm^3$ NaOH	8.60	6.51	11.50

↑
Buffer Solution

How do buffer solutions work?
A solution of ethanoic acid and sodium ethanoate is an example of an acidic buffer, while a solution of ammonia and ammonium chloride is a basic buffer solution.

Consider the following **acid buffer** solution containing $CH_3COOH_{(aq)}$ and the $CH_3COO^-_{(aq)}$ conjugate base:

(i) $CH_3COOH_{(aq)} \rightleftharpoons CH_3COO^-_{(aq)} + H^+_{(aq)}$

(ii) $CH_3COONa_{(aq)} \rightarrow CH_3COO^-_{(aq)} + Na^+_{(aq)}$

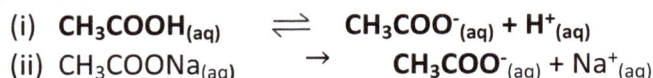

The very soluble salt sodium ethanoate, CH_3COONa is the source of the conjugate base, CH_3COO^-. The Na^+ ion does not take part in the reaction and is referred to as a 'spectator' ion.

When a drop of **strong acid, for example $HCl_{(aq)}$ is added to the buffer solution**, the $H^+_{(aq)}$ ions are removed in the reverse reaction of equation (i) above, as the equilibrium position shifts to the left. The $H^+_{(aq)}$ react with the conjugate base (CH_3COO^-) to form the acid, CH_3COOH.

The following equation represents the removal of H^+ ions in the reverse reaction of equation (i) above: $H^+_{(aq)} + CH_3COO^-_{(aq)} \rightarrow CH_3COOH_{(aq)}$

In accordance with Le Châtelier's Principle, the equilibrium position shifts to the left on increasing the $H^+_{(aq)}$ ion concentration. The $H^+_{(aq)}$ ion concentration decreases and the pH is maintained or changes only very slightly. The salt provides a reservoir of $CH_3COO^-_{(aq)}$ ions which removes the added $H^+_{(aq)}$ ions. When all of the conjugate base has reacted, the solution can no longer act as a buffer solution.

When a drop of **strong base for example $NaOH_{(aq)}$ is added to the buffer solution,** the $OH^-_{(aq)}$ ions are removed, in accordance with Le Châtelier's Principle. The equilibrium position shifts to the right as the ethanoic acid in the forward reaction produces water and the conjugate base $CH_3COO^-_{(aq)}$. The pH of the solution therefore changes only negligibly, if at all.

The following equation represents the removal of the $OH^-_{(aq)}$ ions in the forward reaction of equation (i):

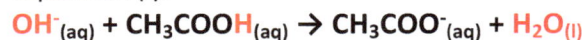

$OH^-_{(aq)} + CH_3COOH_{(aq)} \rightarrow CH_3COO^-_{(aq)} + H_2O_{(l)}$

When all of the ethanoic acid has ionised, the solution can no longer act as a buffer.

Consider the following **basic buffer solution:**

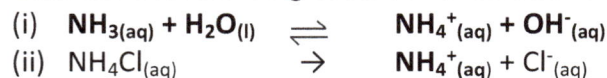

(i) $NH_{3(aq)} + H_2O_{(l)} \rightleftharpoons NH_4^+_{(aq)} + OH^-_{(aq)}$

(ii) $NH_4Cl_{(aq)} \rightarrow NH_4^+_{(aq)} + Cl^-_{(aq)}$

The salt, NH_4Cl, provides a high concentration of the conjugate acid NH_4^+.

On **adding a drop of a strong base** to the above buffer solution, the $OH^-_{(aq)}$ ions are removed in the reverse reaction of equation (i) above, by the conjugate acid, NH_4^+. The equilibrium position shifts to the left, producing $NH_3 + H_2O$, as shown in the following equation:

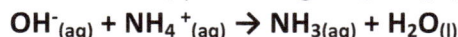

$OH^-_{(aq)} + NH_4^+_{(aq)} \rightarrow NH_{3(aq)} + H_2O_{(l)}$

The pH of the solution therefore changes negligibly.

The solution loses its buffering ability when all of the conjugate acid, NH_4^+ is used up.

On **adding a drop of a strong acid**, the $H^+_{(aq)}$ ions are removed in the forward reaction by the $NH_{3(aq)}$. The equilibrium position therefore shifts to the right in equation (i) above, producing the $NH_4^+_{(aq)}$ ion, as shown in the equation below.

$NH_{3(aq)} + H^+_{(aq)} \rightarrow NH_4^+_{(aq)}$

Some natural buffer solutions include the blood, sea water, ponds, lakes and rivers. They maintain an optimum pH to support life of living organisms. Swimming pools are buffered to prevent algal growth.

The blood's pH at room temperature should be 7.35. If this pH goes down to 7.20, a condition called **acidosis** results, and this requires emergency measures. An overdose of aspirin, uncontrolled diabetes and starvation are some causes of acidosis.

If the pH of the blood drifts upwards towards 7.50, an equally life-threatening condition called **alkalosis** results. Bicarbonate overdose, prolonged hysterics, kidney disease and continual vomiting cause alkalosis.

The Blood's Buffer System

The blood has 3 main buffer systems:
- The hydrogen carbonate buffer: HCO_3^-/ CO_3^{2-}
- The hydrogen phosphate buffer: HPO_4^{2-} /$H_2PO_4^-$ / PO_4^{3-}
- The amino acid buffer

The hydrogencarbonate buffer

Carbon dioxide from respiration produces carbonic acid, H_2CO_3 in cells as follows:

$$CO_{2(g)} + H_2O \rightarrow H_2CO_{3(aq)}$$

Carbonic acid is a weak acid and is in equilibrium with its aqueous ions as follows:

(i) $H_2CO_{3(aq)} \rightleftharpoons H^+_{(aq)} + HCO_3^-{}_{(aq)}$

(ii) $HCO_3^-{}_{(aq)} \rightleftharpoons H^+_{(aq)} + CO_3^{2-}{}_{(aq)}$

In acidic conditions, $H^+_{(aq)}$ ions are removed in the reverse reaction of equation (i) or (ii) above as follows: $H^+_{(aq)} + HCO_3^-{}_{(aq)} \rightarrow H_2CO_{3(aq)}$ OR $H^+_{(aq)} + CO_3^{2-}{}_{(aq)} \rightarrow HCO_3^-{}_{(aq)}$

In basic conditions, the OH^- ions are removed by the acids in the forward reaction as follows:

$$OH^-_{(aq)} + H_2CO_{3(aq)} \rightarrow H_2O_{(l)} + HCO_3^-{}_{(aq)} \quad OR \quad OH^-_{(aq)} + HCO_3^-{}_{(aq)} \rightarrow H_2O_{(l)} + CO_3^{2-}{}_{(aq)}$$

The Amino Acid Buffer

An amino acid has the following general formula:

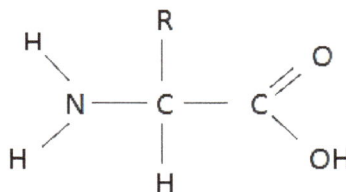

In aqueous solution, the amino acids form **zwitterions.** This occurs when the H^+ from the COOH bonds to the N of the NH_2, as shown below.

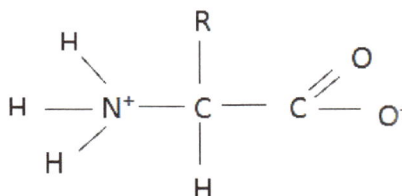

In acid, the $H^+_{(aq)}$ ion bonds to the $COO^-_{(aq)}$ ion of the zwitterion producing a cation as shown:

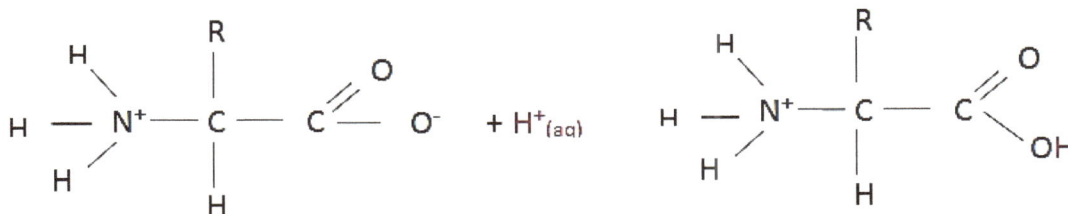

In base, $OH^-_{(aq)}$ ions bond with the H^+ ion to form H_2O and an anion as shown below.

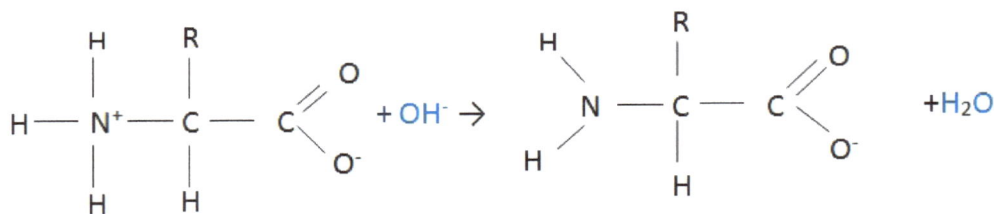

How to Prepare a Buffer solution in the laboratory

1. Decide on the pH of the buffer solution required
2. Select a weak acid/base whose pK_a / pK_b value is $\overset{+}{-}$ 1 the pH of the buffer. For example, if the pH of the buffer is 5.0 then the pK_a of the weak acid used must be between 4 and 6.
   Ka for CH_3COOH is 1.8×10^{-5}
   $pK_a = 4.75$
   Therefore, ethanoic acid would be a suitable weak acid for this buffer.
3. Mix approximately 1 mole of ethanoic acid and approximately 2 moles of sodium ethanoate in 1 dm^3 of solution.
4. Stir thoroughly and measure the pH using a pH meter.

Calculating pH of Buffer solutions

Worked example 11.12

Benzoic acid, C_6H_5COOH is a weak monobasic acid ($K_a = 6.4 \times 10^{-5}$ moldm^{-3}).
a. Calculate the $[H^+_{(aq)}]$ in a 0.02M benzoic acid solution.
b. What is the pH of the solution?
c. Calculate the pH of a solution containing 7.2g of sodium benzoate in 1dm^3 of 0.02M benzoic acid.
d. By how much will the pH change if,
   i) 1cm^3 1.0M NaOH is added to the solution in part (c); (ii) 1cm^3 1.0M HCl is added

a.

$$C_6H_5COOH_{(aq)} \rightleftharpoons C_6H_5COO^-_{(aq)} + H^+_{(aq)}$$

[initial/M]   0.02 $\rightleftharpoons$ 0   0

If y M of
acid ionise   0.02 – y   y   y
at equilibrium

$$Ka = \frac{[C_6H_5COO^-][H^+]}{[C_6H_5COOH_{(aq)}]} \qquad \therefore \quad 6.4 \times 10^{-5} = y^2 / 0.02 - y$$

y $= \sqrt{6.4 \times 10^{-5} \times 0.02}$ ignoring –y

Answer $y = [H^+] = 1.13 \times 10^{-3}$ ɾ

Solution

b. pH = -log₁₀ 1.13 x 10⁻³ = **2.95**

c. $[C_6H_5COONa_{(aq)}] = \dfrac{7.2}{144} = 0.05M$ *(molar mass of C₆H₅COONa = 144)*

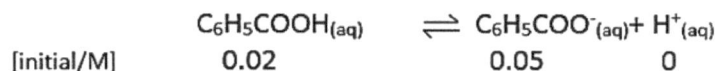

	$C_6H_5COOH_{(aq)}$	$\rightleftharpoons$ $C_6H_5COO^-_{(aq)}$+ H⁺_(aq)	
[initial/M]	0.02	0.05	0
If y M of acid ionises at equilibrium	0.02 − y	0.05 + y	y

$Ka = \dfrac{[C_6H_5COO^-]\,[H^+_{(aq)}]}{[C_6H_5COOH_{(aq)}]}$

$6.4 \times 10^{-5} = \dfrac{(0.05 + y)\,(y)}{0.02 - y}$

$6.4 \times 10^{-5} = \dfrac{0.05\,y}{0.02}$ Ignoring the +y and −y

$y = \dfrac{6 \times 10^{-5} \times 0.02}{0.05}$ = 2.56 x 10⁻⁵ moldm⁻³ = [H⁺]

pH = -log₁₀ 2.5 x 10⁻⁵ = **4.59**

d. i) 1000cm³ NaOH contains 1mol

∴ 1cm³ contains $\dfrac{1}{1000}$ x 1   = **0.001mol OH⁻ ions**

On adding OH⁻_(aq): $C_6H_5COOH_{(aq)} + OH^-_{(aq)} \rightarrow$ $C_6H_5COO^-_{(aq)}$+ H₂O_(l)
Change in concentration   -0.001   +0.001

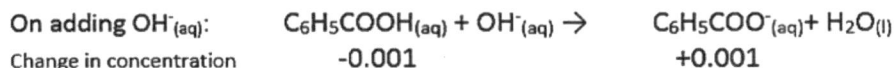

$6.4 \times 10^{-5} = \dfrac{[C_6H_5COO^-]\,[H^+]}{[C_6H_5COOH_{(aq)}]}$ $= \dfrac{(0.05 + y + 0.001)(\,y)}{0.02 - y - 0.001}$

$6.4 \times 10^{-5}\,(0.02 - 0.001) = 0.05 + 0.001\,(y)$ ignoring +y and − y

y = [H⁺] = 2.38 x 10⁻⁵ moldm⁻³

pH = -log₁₀ [H⁺] = -log₁₀ 2.38 x 10⁻⁵ = **4.62.**   The pH increases by **0.03**

d. ii) 1cm³ 1M HCl contains 0.001 mole H⁺ ion.

On adding H⁺_(aq): $C_6H_5COO^-_{(aq)} + H^+_{(aq)} \rightarrow$ $C_6H_5COOH_{(aq)}$

Change in concentration   -0.001   +0.001
$6.4 \times 10^{-5} = = \dfrac{(0.05 + y - 0.001)(\,y)}{0.02 + y + 0.001}$ ignoring +/- y ∴ y=[H⁺]=2.74x10⁻⁵; **pH= 4.56:** pH decreases by

0.03

Buffer systems are important in natural, as well as industrial processes. All pH sensitive reactions, such as enzyme catalysed, biochemical reactions in living organisms must be buffered.

Swimming pools are often buffered to maintain pH conditions that are unfavourable for the growth of algae and other pathogenic organisms.

Buffers are important in the **pharmaceutical, food and textile dyeing industries.**
Medicines are prepared in aqueous solutions which require a constant pH, in order to remain stable and clinically effective.

Buffers are added to foods to maintain a pH that preserves the flavour and appearance of the food. Many food additives act as buffers. For example, sodium citrate added to citric acid containing juices, such as orange juice, creates a buffer solution.

The depth of the colour of dyes, used in the textile industry is related to a narrow pH range, above or below which, the colour imparted to the fabric will vary.

Exercise 11.2
Benzoic acid, C_6H_5COOH, is an organic acid.

i)   i) Write the expression for K_a of benzoic acid. (2 marks)

_____

_____

   ii) Write the equation for the reaction of the conjugate base of benzoic acid and water, and then write the expression for K_b of the conjugate base. (2 marks)

_____

_____

b. pK_a of HCN is 9.21 and that of HF is 3.17. Which of the two, CN^- and F^- is the stronger base? Explain your answer. (3 marks)

_____

_____

_____

c. i) Write appropriate equations to illustrate how the pair of compounds, NaH_2PO_4 and Na_2HPO_4 can act as a buffer inside body cells. (2 marks)

_____

_____

   ii) Calculate the pH of a buffer solution containing 1 M NH_4Cl and 0.1 M NH_3 solution, and comment on the change in the pH value.
   (4 marks)

Calculate the pH of the buffer solution when 0.5 cm³ of 1 M H_2SO_4 is added to it.

(3 marks)

11.3
a. i) State Le Châtelier's principle (1 mark)

b. State how the equilibrium position, composition and equilibrium constant for the following reaction: $CH_{4(g)}$ + $2H_2S_{(g)}$ $\rightleftharpoons$ $CS_{2(g)}$ + $2H_{2(g)}$ ΔH +ve, will be affected by:
i) The addition of CH_4 gas (2 marks)

ii) The addition of CS_2 (2 marks)

iii) A decrease in the volume of the container (2 marks)

iv) An increase in temperature (3 marks)

Solubility, Solubility product and the Common ion effect

Solubility refers to the property of a solid, liquid or gas called a **solute**, to dissolve in a solid, liquid, or gas called a **solvent**. Water is a universal solvent dissolving a large number of chemical substances.

The solubility of a solid in water is **defined** as the mass of solute that dissolves in 100 g of water. The units of solubility may be expressed as g dm^{-3} or mol dm^{-3} of solution.

Factors that affect the solubility of a solute are temperature, the nature of the solvent, polar or non-polar and pressure, for gases.

Increased temperature increases the solubility of solids but decreases the solubility of gases.

Polar solutes will dissolve in polar solvents and non-polar ones dissolve in **non-polar** solvents.

The solubility of gases increases with increased **pressure.**

Many solids are only partially soluble in water, and at a given temperature an equilibrium exists between undissolved solid particles and its aqueous ions. The **solubility product** applies, where solids such as calcium hydroxide is only partially soluble in water, forming a suspension. The following equation shows the equilibrium between solid calcium hydroxide and its ions.

$$Ca(OH)_{2(s)} \rightleftharpoons Ca^{2+}_{(aq)} + 2OH^-_{(aq)}$$

At equilibrium, the product of the concentration of the aqueous ions raised to their coefficient in the balanced equation is constant. This constant is called the **solubility product constant, K$_{sp}$.**

For the above equation, the solubility product constant expression is as follows:

$$K_{sp} = [Ca^{2+}_{(aq)}][OH^-_{(aq)}]^2$$

K_{sp} is a measure of the solubility of a solid. The higher K_{sp} is, the more soluble the solid.

compound	K$_{sp}$ at 25°C
Ca(OH)$_2$	6.5×10^{-6}
CaCO$_3$	4.5×10^{-9}
AgCl	1.8×10^{-10}
AgI	8.3×10^{-17}
CaSO$_4$	2.4×10^{-5}
BaSO$_4$	1.08×10^{-10}
Al(OH)$_3$	3.0×10^{-34}

Worked example 11.13

Calculate the concentration of the aqueous ions in a saturated solution of barium sulphate.

$$BaSO_{4(s)} \rightleftharpoons Ba^{2+}_{(aq)} + SO_4^{2-}_{(aq)}$$

Let the solubility of BaSO$_4$ be x, then at equilibrium, [ions]/M is   x   x

$$K_{sp} = [Ba^{2+}_{(aq)}][SO_4^{2-}_{(aq)}]$$
$$1.08 \times 10^{-10} = x^2$$
$$\therefore x = \textbf{1.04} \times \textbf{10}^{-4} \textbf{ mol dm}^{-3}$$

Worked example 11.14

Write an expression for the solubility product constant, K_{sp} for calcium phosphate and determine the unit.

$$Ca_3(PO_4)_{2(s)} \rightleftharpoons 3Ca^{2+}_{(aq)} + 2PO_4^{3-}_{(aq)}$$

K_{sp} = $[PO_4^{3-}_{(aq)}]^2 [Ca^{2+}_{(aq)}]^3$
Unit = $(mol\ dm^{-3})^2 (mol\ dm^{-3})^3$
$mol^5\ dm^{-15}$

Exercise 11.4

a. Calculate the solubility of lead iodide in water at 25°C, given that K_{sp} at this temperature is $7.1 \times 10^{-9}\ mol^3\ dm^{-9}$.

b. The solubility of lead chloride at a given temperature is $0.0125\ mol\ dm^{-3}$. Calculate K_{sp} for lead chloride at that temperature.

The common ion effect

Consider a suspension of barium sulphate, in which equilibrium exists between undissolved solid and its aqueous ions. If about 1 cm³ of **barium** nitrate is added, according to Le Chatilier's Principle, the equilibrium position shifts to the left, as sulphate ions in solution react with added barium ions in order to lower their increased concentration. This results in barium sulphate precipitating out of solution, and its solubility therefore decreasing. The solubility of barium sulphate would also decrease if sodium **sulphate solution** is added to the suspension.

The effect of decreasing the solubility of a sparingly soluble solid, on adding a **common ion**, is called **the common ion effect**.

Worked example 11.15

0.01 mol of Ba^{2+} ions are added to a suspension of barium sulphate at 25°C. Calculate the solubility of the barium sulphate.

$$BaSO_{4(s)} \rightleftharpoons Ba^{2+}_{(aq)} + SO_4^{2-}_{(aq)}$$

Let the solubility of $BaSO_4$ be x, then at equilibrium, [ions]/M is $(x + 0.01) + x$
K_{sp} = $[Ba^{2+}_{(aq)}] [SO_4^{2-}_{(aq)}]$ = $(0.01 + x)(x)$ (ignore '+x' mathematically)
1.08×10^{-10} = $(0.01x)$ ∴x = $1.08 \times 10^{-10} / 1 \times 10^{-2}$ = **1.08 x10⁻⁸ mol dm⁻³**
The solubility has decreased from $1.04 \times 10^{-4}\ mol\ dm^{-3}$ to 1.08 x10⁻⁸ mol dm⁻³.

Ksp for a partially soluble solid like Ca(OH)₂ can be determined in the laboratory as follows:

Make a saturated solution of calcium hydroxide, by shaking about 25 g of the solid with 100 cm³ distilled water in a conical flask. Leave the mixture overnight so that equilibrium can be established between undissolved solid and its aqueous ions. Record the temperature of the liquid.

Decant the aqueous layer of the suspension, leaving the undissolved solid behind. Titrate 25 cm³ of this solution with a standard solution of an acid, for example, 0.1 mol dm⁻³ $HCl_{(aq)}$.

Calculate the $OH^-_{(aq)}$ ion concentration.

If $[OH^-_{(aq)}] = y$, then $[Ca^{2+}_{(aq)}] = \frac{1}{2} y$ according to the equation: $Ca(OH)_{2(s)} \rightleftharpoons Ca^{2+}_{(aq)} + 2 OH^-_{(aq)}$

$Ksp = [Ca^{2+}_{(aq)}] [OH^-_{(aq)}]^2 = \frac{1}{2} y \times y^2$

Kidney stones are solid calcium oxalate, CaC_2O_4, or calcium carbonate, $CaCO_3$, formed in the calyx of the kidney. The following equation represents the equilibrium reaction:

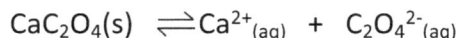

$CaC_2O_4(s) \rightleftharpoons Ca^{2+}_{(aq)} + C_2O_4^{2-}_{(aq)}$

High dietary intake of calcium or oxalate ions, shift the equilibrium position to the left, precipitating out solid CaC_2O_4 which forms 'stones'. These become painful when they pass down the uréter. They may require surgery, if they are lodged, causing obstruction in the uréter.

Exercise 11.5

a. Calculate the solubility in g dm⁻³ of $CaCO_3$ in:

   i)   pure water

   ii) 0.0250 mol dm⁻³ sodium carbonate

   iii) 0.150 mol dm⁻³ calcium nitrate

Chapter 12
REACTION KINETICS

Rates of Reactions

The rate of a chemical reaction is defined as the change in the concentration of a reactant or product with changing time.

$$Rate = -\frac{\Delta \text{[reactant]}}{\Delta \text{time}}$$

$$= +\frac{\Delta \text{[product]}}{\Delta \text{time}}$$

The unit of rate: $moldm^{-3}s^{-1}$

Rate and time are inversely related:
Fast rate = short time
Slow rate = long time

$$Rate \propto \frac{1}{\text{time}}$$

A fast rate of reaction is desirable in some processes and a slow rate in others, as shown in the table below.

Fast	Slow
For medications to act	Preservation of food
For metabolic reactions	Rusting of products made from iron
In industry (Haber Process)	

Measuring rates of reactions experimentally

The following are four techniques that can be used in measuring the rate of a chemical reaction.

1. Titrimetric analysis
2. Colorimetric analysis
3. Pressure measurements
4. Conductimetric analysis

Titrimetric analysis

This method is most suitable for reactions in aqueous solution. It gives a direct measurement of the reaction rate.

Consider an experiment in which the rate of decomposition of hydrogen peroxide is being measured.

$$2H_2O_{2(l)} \longrightarrow 2H_2O_{(l)} + O_{2(g}$$

Procedure
1. Obtain 5 conical flasks.
2. Pipette $25cm^3$ of 1.0M $H_2O_{2(aq)}$ in each flask.
3. To the first flask, add 1g of a suitable catalyst and simultaneously start a stop clock.

4. After 2 minutes, quench the reaction, that is, stop the reaction with a suitable chemical and titrate with a suitable standard solution, for example acidified potassium permanganate solution.

5. Calculate the concentration of the H_2O_2 remaining after 2 minutes. Repeat all the above steps for the reactions in the other flasks, stopped after 4, 6, 8, and 10 minutes respectively.

6. Tabulate the results in a table with time and $[H_2O_2]/moldm^{-3}$.

Colorimetric analysis

This technique is based on changes in the intensity of the colour of a reactant or product during the course of a reaction. Colour changes indicate changes in the concentration of reactant or product with change in time. The instrument used is the photoelectric colorimeter. The method works on the principle that the more concentrated a coloured substance, the darker its colour and the greater its absorption of visible light. The less concentrated the substance, the paler its colour and the lower its absorption of light. The absorbance of visible light by a colourless solution is zero.

Following is a study of the reaction between bromine which is coloured, and methanoic acid. The reaction is catalysed by an acid.

$$Br_{2(aq)} + HCOOH_{(aq)} + H^+_{(aq)} \longrightarrow 2\ Br^-_{(aq)} + 3H^+_{(aq)} + CO_{2\ (g)}$$

Procedure

1. Place about 1 cm^3 of 1 moldm^{-3} reddish brown bromine, in the small container called a cuvette.

2. Place it in the cuvette holder of the instrument.

3. Switch on the light source and record the initial absorbance reading from the instrument.

4. Add an excess of methanoic acid while simultaneously starting a stop clock.

5. Record the absorbance reading at regular 10 second intervals for example.

As time progresses, the concentration of Br_2 decreases, the colour fades and the absorbance reading decreases. The absorbance values are proportional to the concentration of Br_2 remaining at each time interval. If the product was coloured and the reactant colourless, then the absorbance increases as the concentration and colour intensity of the product increases.

Pressure measurements

This technique is suitable for gaseous reaction systems.

This involves changes in pressure during the course of the reaction. An example is monitoring the pressure of CO_2 produced when a carbonate such as $CaCO_3$ is thermally decomposed.

$$CaCO_{3(s)} \longrightarrow CaO_{(s)} + CO_{2(g)}$$

The pressure of the gas produced can be recorded at regular time intervals, and a graph of pressure of product verses time, plotted. The pressure of the gaseous product is directly proportional to its concentration.

Conductimetric Analysis

Many reactions in aqueous solution, involve changes in the concentration of the ions of the reactant or product, as the reaction proceeds. The electrical conductivity of the solution therefore changes with time.

This technique involves inserting two inert electrodes into the reaction mixture and monitoring the change in electrical conductivity with time.

For example, the reaction between calcium and sulphuric acid to produce calcium sulphate and Hydrogen is suitable. Only one reactant, sulphuric acid is aqueous. The decrease in the concentration of its ions and therefore its electrical conductivity can be followed and recorded using an electrochemical cell.

$$Ca_{(s)} + H_2SO_{4(aq)} \rightarrow CaSO_{4(s)} + H_{2(g)}$$

Factors Affecting the Rates of Chemical Reactions

Many factors influence the rate at which chemical reactions proceed. These include:

1. **Concentration of reactants**
2. **Pressure, if gaseous**
3. **Temperature**
4. **Surface area of a solid**
5. **Presence of a catalyst**

The Effect of Concentration on Reaction Rate

Analysis of Raw Concentration Data from Titrimetric Analysis

Consider the following results for the catalytic decomposition of hydrogen peroxide by titrimetric analysis.

Time/min	0	2	4	6	8	10
$[H_2O_2]$/M	1.00	0.49	0.26	0.20	0.10	0.06

These results can be analysed as shown below, in order to determine the relationship between the concentration of hydrogen peroxide and the rate of the reaction

Analysis of data

1. **Plot a graph of [reactant] versus time,** as shown in Fig. 12a.

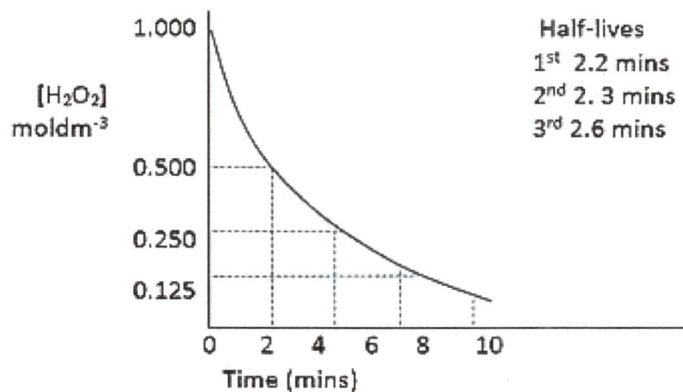

Half-lives
1st 2.2 mins
2nd 2. 3 mins
3rd 2.6 mins

Fig. 12a *graph of [H$_2$O$_2$] versus time*

2. Plot a graph of Rate versus [Reactant]

$$Rate = \frac{\Delta\,[reactant]}{\Delta time} = \frac{y}{x} = gradient$$

This graph, Fig. 12b, shows a straight line passing through the origin. Therefore, the rate is directly proportional to the concentration of the reactant. $[H_2O_2]$ when the rate of a reaction is directly proportional to the concentration of the reactant, the reaction is said to be **first order** with respect to that reactant.

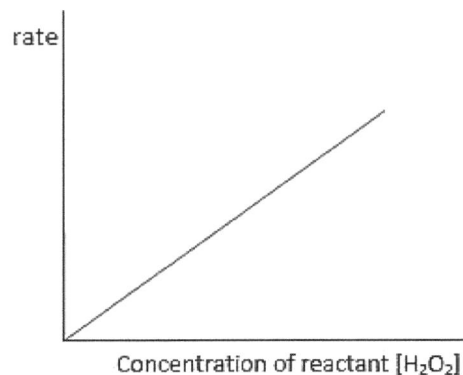

rate

Concentration of reactant $[H_2O_2]$

Fig. 12b *rate vs concentration*

3. The **half-life** of the reactant can also be determined from the graph of concentration against time. Fig. 12a shows three half-lives of 2.2 minutes, 2.3 minutes and 2.6 minutes. The half-life is constant given experimental error.

The half-life of first order reactions is independent of the initial concentration of the reactant. It therefore remains constant and is given by:

$$t_{1/2} = \frac{\ln 2}{k} = \frac{0.693}{k}$$
(k is the rate constant)

Order of reaction and rate equations

The order of a reaction describes how the rate of a reaction is affected by the concentration of the reactants.

The order of a reaction is defined as the sum of the powers to which the concentration of each reactant is raised, in an experimentally determined rate equation.

The order of a reaction with respect to each reactant is defined as the power to which the concentration of that reactant is raised, in the rate equation.

Fig. 12b shows that the rate of reaction is directly proportional to the concentration of hydrogen peroxide. That is: Rate α $[H_2O_2]$ ∴ Rate = k $[H_2O_2]$ (**rate equation**)

The reaction is said to be **first order** with respect to hydrogen peroxide.

The rate equation or rate law is a mathematical expression which shows the relationship between the rate of a chemical reaction and each or all of the reactants incorporating a rate (velocity) constant.

Reactions in which the rate of reaction is independent of the concentration of the reactant are said to be **zeroth order** with respect to the reactant. Figures 12c and 12d demonstrate this. When the concentration of the reactant is doubled or tripled, the rate of the reaction remains unchanged.

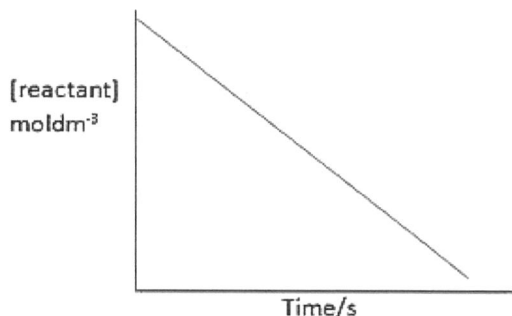

Fig. 12c *concentration versus time for a zeroth order reaction*

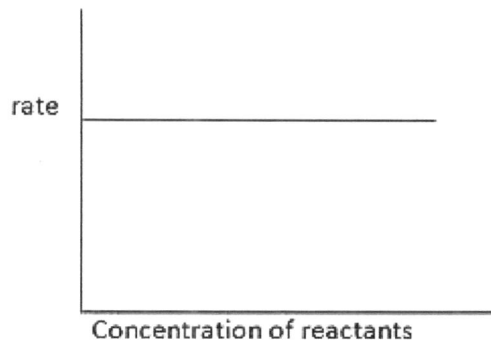

Fig. 12d *rate vs. concentration for a zenoth order reaction*

Figures 12e and 12f below are graphs that are characteristic of **second order reactions.**

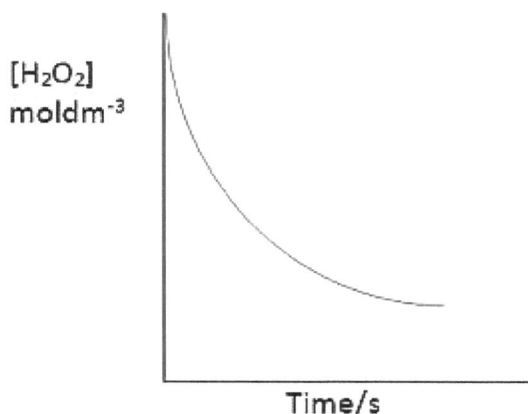

Fig. 12e *concentration versus time for second order reaction*

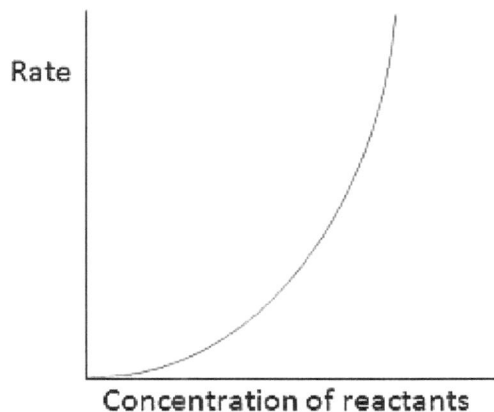

Fig. 12f *rate versus concentration versus for a second order reaction*

A second order reaction is one in which the rate is proportional to the square of the concentration of a reactant.

For 2nd order reactions, the half-life is dependent on the initial reactant concentration. The half-life therefore changes with time, as the concentration of the reactant decreases. It is given by:

$$t_{1/2} = \frac{1}{[initial\ reactant]\ k}$$

Consider the following equation: A + B → C

If from experimental data:

Rate α [A] then the Rate = k [A]

The reaction is therefore 1st order with respect to A, since the power to which the concentration of A is raised is 1.

If Rate α $[B]^2$ then the Rate = k $[B]^2$

The reaction is therefore 2nd order with respect to B since the power to which the concentration of B is raised is 2.

The Rate equation for the overall reaction is therefore as follows:

Rate = k [A] $[B]^2$

The order of the reaction with respect to reactant A is 1 and with respect to B, is 2

Hence the overall order of the reaction is 1 + 2 = 3.

In the rate equation, k is called the **velocity constant or the rate constant**. It is defined as the proportionality constant in a rate equation. It is constant at a constant temperature. The use of a catalyst and a change in temperature change the rate constant.

Analysis of initial rates data

Consider the reaction between hydrogen and nitrogen monoxide:

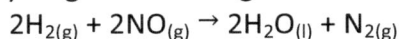

$$2H_{2(g)} + 2NO_{(g)} \rightarrow 2H_2O_{(l)} + N_{2(g)}$$

The following is a table of experimental results.

Experiment No.	[Initial NO] moldm^{-3}	Initial [H$_2$] moldm^{-3}	Initial rate moldm^{-3}
1	8×10^{-3}	1×10^{-3}	3×10^{-3}
2	8×10^{-3}	2×10^{-3}	6×10^{-3}
3	8×10^{-3}	3×10^{-3}	9×10^{-3}
4	1×10^{-3}	8×10^{-3}	0.5×10^{-3}
5	2×10^{-3}	8×10^{-3}	2.0×10^{-3}
6	3×10^{-3}	8×10^{-3}	4.5×10^{-3}

Table 12.1 *experimental results*

In experiments 1, 2 and 3 where the concentration of NO is constant, when the concentration of hydrogen is doubled, the rate is doubled also. When the concentration of hydrogen is tripled, the rate is tripled also. This shows that the rate of the reaction is directly proportional to the concentration of hydrogen.

> Rate α [H$_2$]
> Rate = k[H$_2$]

The reaction is therefore 1st order with respect to hydrogen.

In experiments 4, 5 and 6 where the concentration of hydrogen is kept constant, and the concentration of nitrogen monoxide is doubled, the reaction rate increased fourfold and when the concentration was tripled, the rate increased nine-fold. This shows that the rate is proportional to the square of the concentration of nitrogen monoxide.

> Rate α [NO]^2
> Rate = k[NO]^2

Therefore, the reaction is 2nd order with respect to nitrogen monoxide.

The rate equation for the reaction is as follows:

> Rate = k[H$_2$] [NO]^2 The overall order of the reaction is therefore 3 or 3rd order.

The following is an **equation** that can be used to calculate order of a reaction:

$$\frac{\text{Rate in experiment 2}}{\text{Rate in experiment 1}} = \left(\frac{\text{[reactant] in experiment 2}}{\text{[reactant] in experiment 1}}\right)^n$$

Where experiments '1' and '2' refer to any two experiments under consideration
For example, from the table above in experiments 4 and 6:

$$\frac{4.5 \times 10^{-3}}{.5 \times 10^{-3}} = \left(\frac{.003}{.001}\right)^n$$

$$\therefore \quad 9 = (3)^n$$

$$n = 2$$

Therefore, this reaction is 2nd order with respect to NO.

The Effect of Temperature on Reaction Rate

The effect of temperature on the rate of different types of chemical reactions is shown in the following diagrams. Figures 12g, h and i show the effect of temperature on a) most chemical reactions, b) explosive reactions and c) enzyme catalysed reactions, respectively.

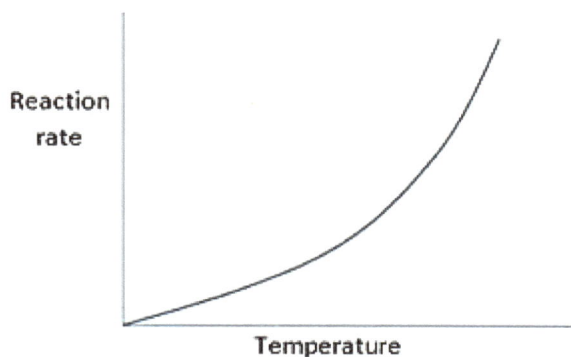

Fig. 12g *most chemical reactions*

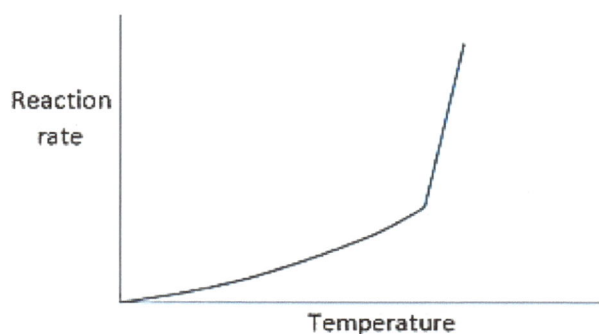

Fig. 12h *explosive reactions*

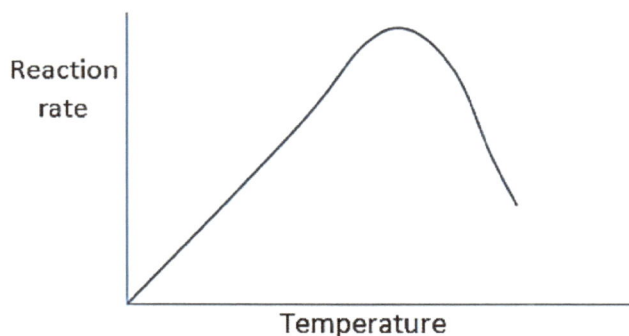

Fig. 12i *enzyme catalysed reactions*

During the course of chemical reactions, reactant particles in the gas phase, are in continuous motion, colliding with each other.

A collision is said to be an **'effective collision'** or a successful collision, if old bonds are broken and new bonds are made. Most of the collisions of particles are not effective collisions. A collision can only be effective if the particles have the energy of activation and are appropriately aligned.

The following diagrams illustrate different possible alignment of the reactant particles in a chemical reaction.

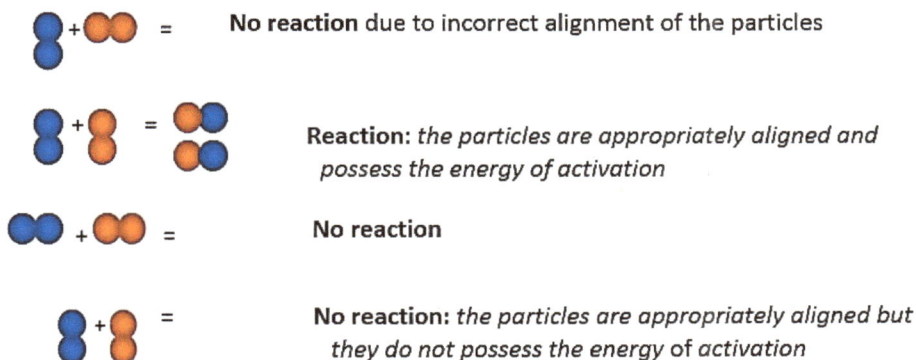

 No reaction due to incorrect alignment of the particles

 Reaction: *the particles are appropriately aligned and possess the energy of activation*

 No reaction

 No reaction: *the particles are appropriately aligned but they do not possess the energy of activation*

In a chemical reaction system, the reactant particles have varying amounts of kinetic energy amongst them.

This is illustrated in the **Maxwell/Boltzman distribution curve** in Figure 12j.

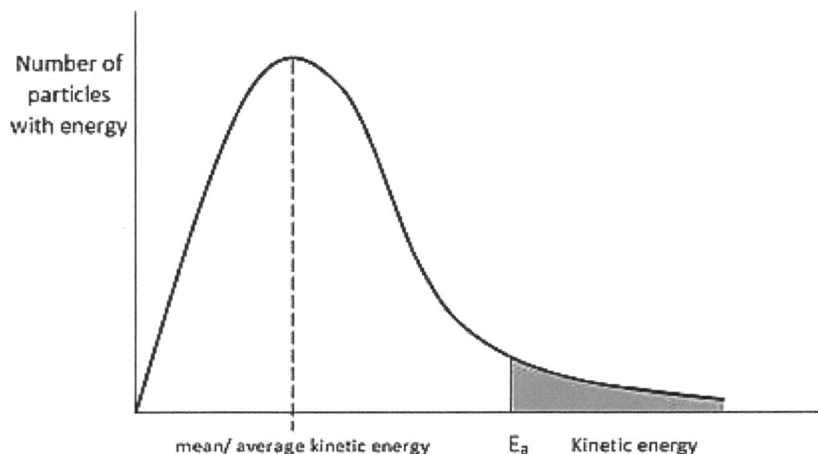

Fig. 12j *Maxwell/Boltzman energy distribution curve*

The curve represents the varying amounts of kinetic energy possessed by the particles of a reactant at any given temperature. It shows that only a few particles have very low kinetic energy and very high kinetic energy at any temperature. The energy represented by the peak is the average kinetic energy of the particles at that temperature. For reactant particles to experience **successful** or **effective collisions**, they must possess a minimum amount of energy, the **energy of activation (E_a)**.

Figure 12k shows the energy distribution of reactant particles at temperature T1 and at temperature T2. Temperature T2 is 10 degrees higher than T1.

At temperature T2, fewer particles have low kinetic energy and there is an increase in the average kinetic energy of the particles. The number of particles with the energy of activation and more, also increases and the frequency of effective collisions therefore increases. This results in an increase in the rate of reaction.

Only the particles that possess the energy of activation and more will produce effective collisions and form products. The shaded area under T1 in Figure 12k, represents the total number of particles that will react at that temperature, per unit time. However, when the temperature is increased by 10 degrees, as shown in T2, note that a larger number of particles in the area that is lined will have effective collisions, and forms product. At this increased temperature, the frequency of effective collisions will therefore increase, hence the rate of reaction will increase.

In general, the rate of a chemical reaction doubles for every 10 degree rise in temperature.

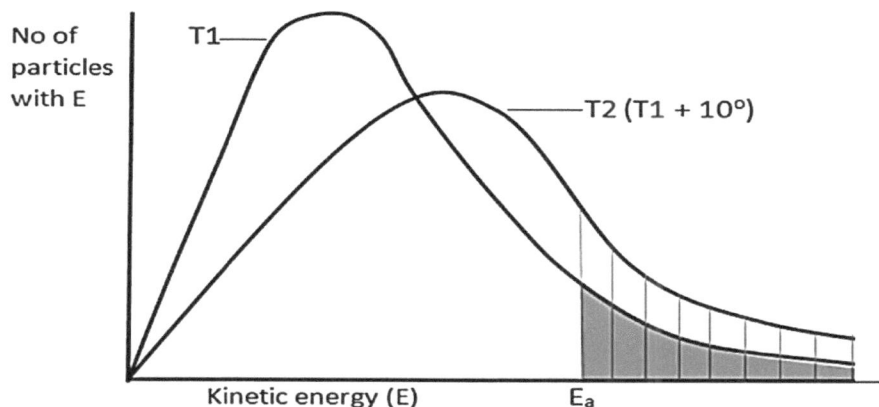

Fig. 12k *The Maxwell/Boltzman distribution curves at increased temperature*

The Effect of Catalysts on Reaction Rate

There are two types of catalysts or catalytic reactions (catalysis):

1. Homogenous catalysis
2. Heterogenous catalysis

In **homogenous** catalysis, the catalyst and the reactants and products, are in the same physical state. In **heterogenous** catalysis, the catalyst and the reactants and products are in different physical states.

Positive catalysts increase the rates of reaction by lowering the energy of activation, resulting in more particles experiencing effective collisions per unit time. This causes the rate of the reaction to increase. **Negative catalysts** decrease the rate of reaction by increasing the energy of activation.

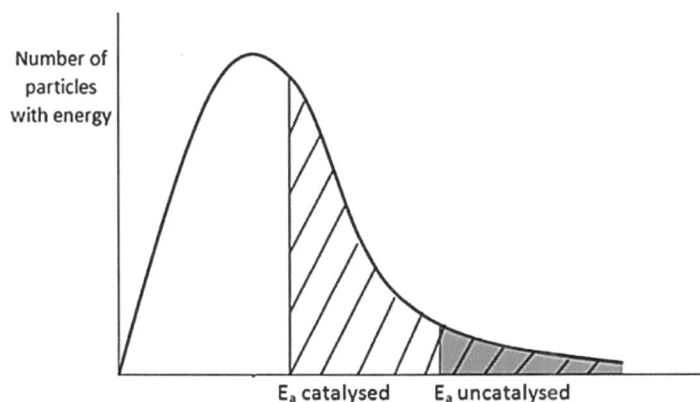

Fig. 12L *Maxwell/Boltzman curve for a catalysed reaction*

The Maxwell/Boltmzann Curve in Fig. 12L above, shows that at a given temperature, only those particles that are shaded have the energy of activation (E_a) and will produce effective collisions. In the presence of a positive catalyst, however, the energy of activation is lowered and a larger number of particles now have the energy of activation and more as indicated in the lined area of the diagram. This results in an increase in the frequency of effective collisions and an increase in the rate of reaction.

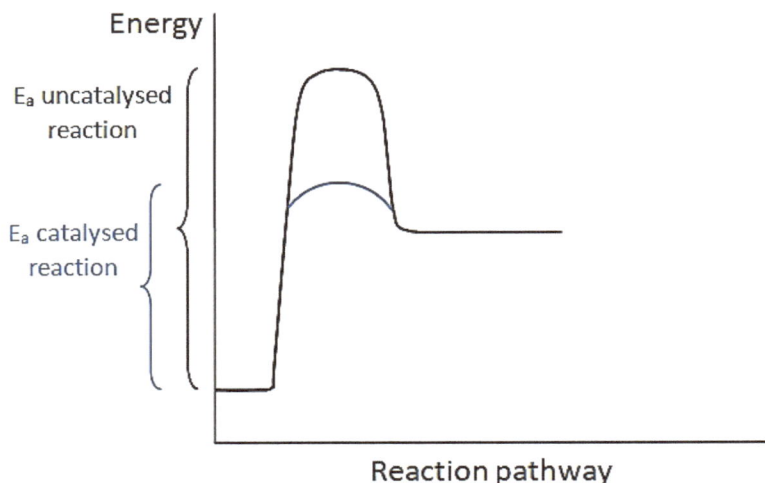

Fig. 12m *energy profile diagram for a catalysed and uncatalysed reaction*

How catalysts work

Heterogenous catalysis

Consider the Haber Process for the manufacture of ammonia:

$$N_{2(g)} + 3H_{2(g)} \rightleftharpoons 2NH_{3(g)}$$

This reaction takes place on the surface of ***finely divided iron*** as a catalyst.

There are three main steps involved in this catalytic reaction.

1. Adsorption
2. Reactivity
3. Desorption

During **adsorption** the reactant particles are bonded to the surface of the catalyst. This bonding process releases energy, which results in a decrease in the energy of activation needed.

Also, on the surface of the catalyst, the reactant particles are concentrated.

The frequency of effective collisions on the catalytic surface increases and the **reaction occurs at a very fast rate**. The product formed is released from the surface of the catalyst. This process is called **desorption**.

The surface of the catalyst is therefore made available for further adsorption and reactivity.

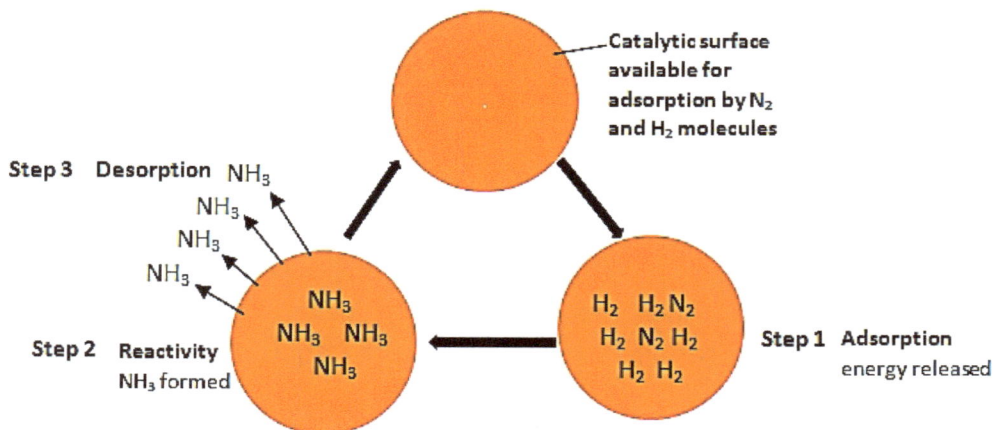

Fig. 12n *diagram showing the three stages in heterogenous catalysis*

Catalytic converters placed in the engines of vehicles use heterogenous catalysis to catalyse the conversion of air-polluting gases to less polluting products.
For example,

$$SO_{2(g)} + NO_{2(g)} \rightarrow SO_{3(g)} + NO_{(g)}$$
$$NO_{(g)} + \tfrac{1}{2}O_{2(g)} \rightarrow NO_{2(g)}$$

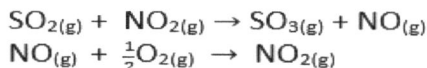

Many heterogenous catalysts are "poisoned" by substances that will coat their surfaces and prevent adsorption, For example, dust. Lead in leaded gasoline poisons the catalysts in catalytic converters.

Homogenous Catalysis

Homogenous Catalysts provide alternate reaction pathways with lower energy of activation.
For example, consider:

$A + B \rightarrow AB$,   Suppose that the energy of activation, E_a = 1960 kJ for the uncatalysed reaction.

In the presence of a homogenous catalyst, 'C':

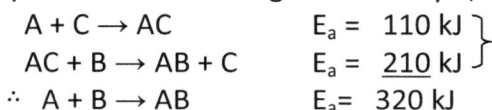

$A + C \rightarrow AC$	E_a = 110 kJ
$AC + B \rightarrow AB + C$	E_a = 210 kJ
$\therefore A + B \rightarrow AB$	E_a= 320 kJ

The Contact process for the manufacture of H_2SO_4 is catalysed by a heterogenous catalyst. However, the manufacture of sulphuric acid was once catalysed using a homogenous catalyst, in the Lead chamber process. NO_2 was the homogenous catalyst used as shown below.

$$2SO_{2(g)} + O_{2(g)} \rightleftharpoons 2SO_{3(g)}$$

In the presence of the homogenous catalyst, NO_2:

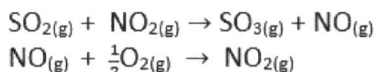

$$SO_{2(g)} + NO_{2(g)} \rightarrow SO_{3(g)} + NO_{(g)}$$
$$NO_{(g)} + \tfrac{1}{2}O_{2(g)} \rightarrow NO_{2(g)}$$

Catalysts are not used up but are regenerated at the end of the reaction.
Enzymes are catalysts that speed up the rate of biochemical reactions.
These are sensitive to both temperature and pH because they are protein in nature. There is an optimum temperature and optimum pH at which enzymes function best. They are destroyed at high temperatures and at pH values that are too high or too low.

Reaction Mechanism

Most chemical reactions take place in a series of simple steps. This series of simple reaction steps is known as the **reaction mechanism**. The slowest step in the series is the **rate determining step**. The mole ratio, in which the reactants in the slowest step combine, gives the order of the reaction with respect to each reactant. Therefore the overall order of the reaction can be deduced.

Worked example 12.1
Predict a possible reaction mechanism for the following reaction:

$4HI + O_2 \rightarrow 2H_2O + 2I_2$ given that, Rate = k[HI] [O_2]

Solution
Possible steps are as follows:

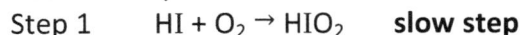

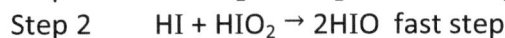

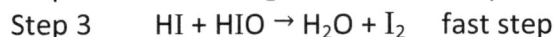

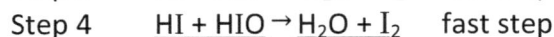

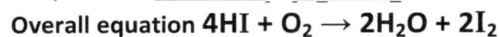

Step 1 $HI + O_2 \rightarrow HIO_2$ **slow step**
Step 2 $HI + HIO_2 \rightarrow 2HIO$ fast step
Step 3 $HI + HIO \rightarrow H_2O + I_2$ fast step
Step 4 $\underline{HI + HIO \rightarrow H_2O + I_2}$ fast step
Overall equation $4HI + O_2 \rightarrow 2H_2O + 2I_2$

Note *that in the slow step, 1 mole of HI reacts with 1 mole of oxygen. Therefore, the reaction is 1st order with respect to HI and 1st order with respect to oxygen. The overall reaction is 2nd order.*

Worked example 12.2
Predict a possible reaction mechanism for the reaction below:
$2NO_2 \rightarrow 2NO + O_2$

The reaction is 2nd order with respect to the NO_2.

Solution
Rate $= k[NO_2]^2$

$NO_2 + NO_2 \rightarrow NO_3 + NO$ **slow step**
$\underline{NO_3 + NO \rightarrow O_2 + 2NO}$ fast step
$2NO_2 \rightarrow O_2 + 2NO$

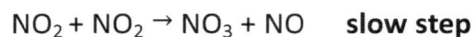

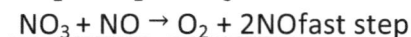

The rate equation for any given reaction can be deduced from knowledge of the reaction mechanism, since the slow step reflects the rate equation.

That is, if the slow step is: $2A+B \longrightarrow A_2B$, then the reaction is second order with respect to reactant A, and first order with respect to reactant B. Therefore, the rate equation is as follows: Rate $= k[A]^2[B]$.

Importance of reaction rate studies

In industry, the efficient use of time and resources is essential. Rate studies give information about these. Chemists, other scientists and industrialists aim to obtain maximum products from the minimum amount of raw materials, in the shortest possible time. Thus they save on fuel costs and are able to make maximum profit.

If it is known how the amount of each raw material in a chemical process affects the rate and if the steps are also known, especially the rate determining step, then wastage of raw materials and time can be avoided.

Straight line graphs plotted against time for zeroth, first and second order reactions

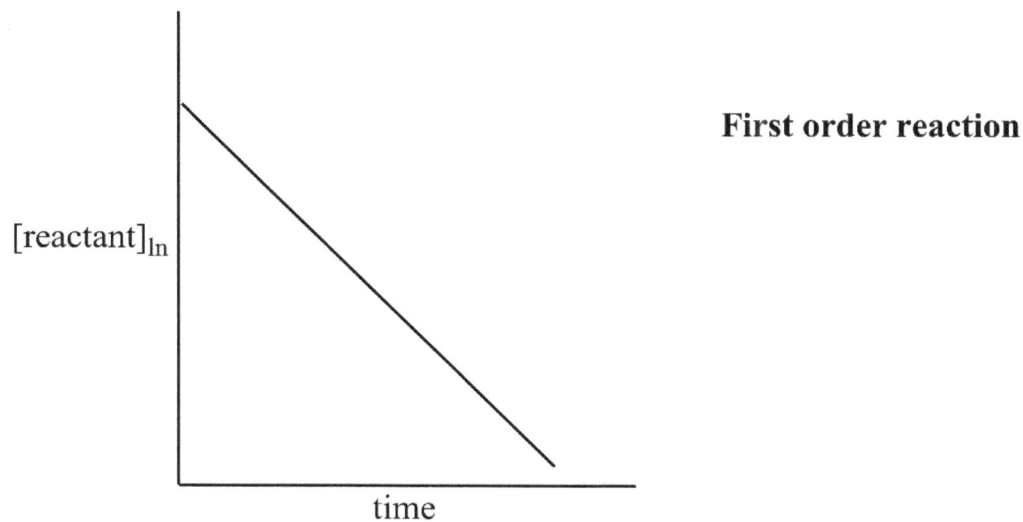

First order reaction

[reactant]$_{\ln}$

time

Fig. 12o *Straight line plot for a **first order reaction***

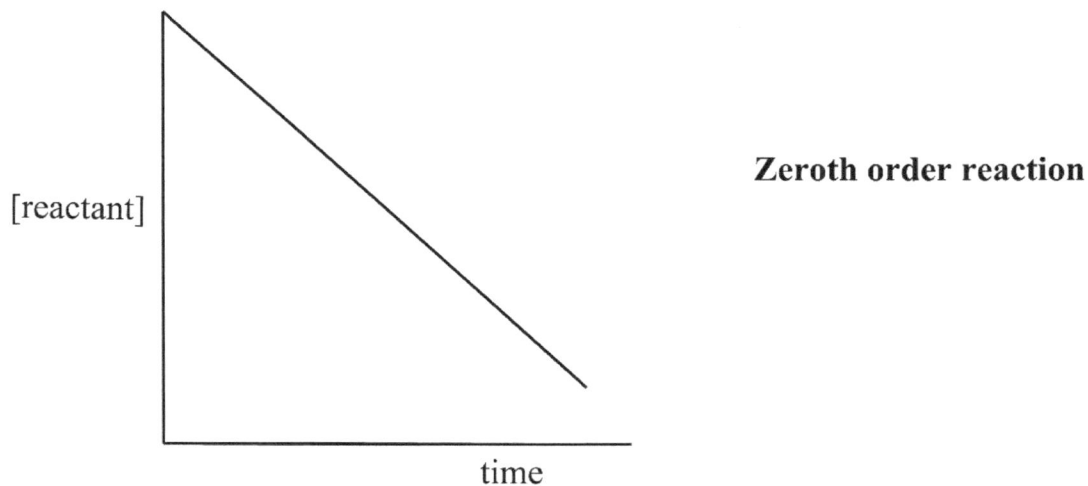

Zeroth order reaction

[reactant]

time

Fig. 12p *Straight line plot for a **zeroth order reaction***

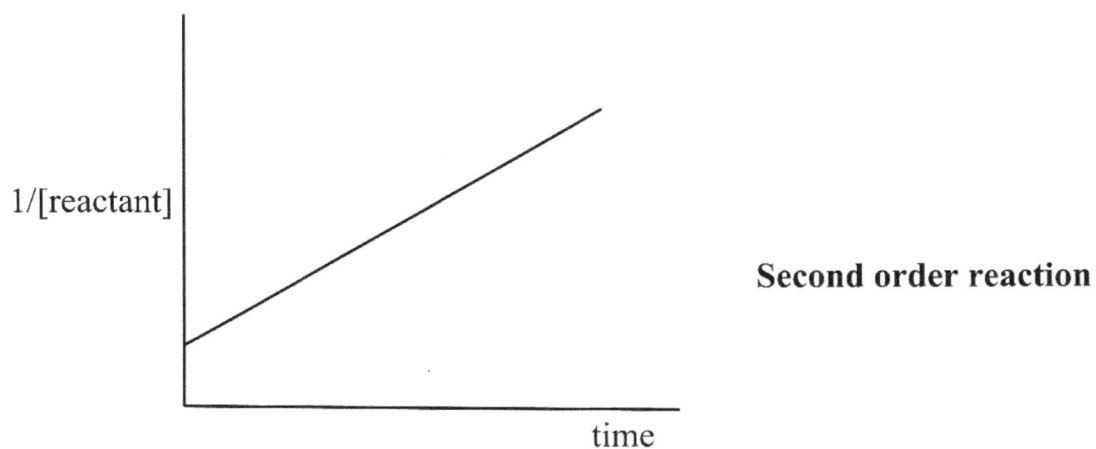

1/[reactant]

Second order reaction

time

Fig. 12q *Straight line plot for a **second order reaction***

Exercise 12.1

Consider the following data:

Two gases, X_2 and Y_2, react according to the equation; $X_{2(g)} + Y_{2(g)} \rightarrow 2XY_{(g)}$
Experiments to determine the order of this reaction gave the following results:

Experiment No.	Initial[X] moldm^{-3}	Initial [Y] moldm^{-3}	Initial rate moldm^{-3} s^{-1}
1	0.15	0.15	1 x 10^{-4}
2	0.15	0.30	4 x 10^{-4}
3	0.15	0.45	9 x 10^{-4}
4	0.30	0.15	1 x 10^{-4}
5	0.45	0.15	1 x 10^{-4}

a. Deduce the order of this reaction with respect to i) X and ii) Y (3 marks)

_____

_____

_____

_____

b. Write a rate equation for the reaction of X with Y. (1 mark)

_____

c. Using the results from experiment 1, calculate a value for the rate constant k. (3 marks)

d. What is the unit of k? (1 mark) _____

e. Predict a possible reaction mechanism for the following reaction:
   $2NO_2Cl \rightarrow 2NO_2 + Cl_2$ given that Rate = k[NO_2Cl] (2 marks)

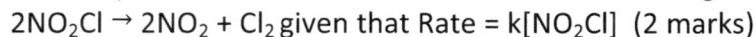

_____

_____

_____

_____

12.2.

a. Define the following terms:

   ii)   Rate of reaction (2 marks)_____

_____

   iii)   Order of reaction (2 marks)

_____

_____

   iii)   Half-life (2 marks)

_____

_____

      iv) ^{131}I decomposes by a first order reaction. The half-life of ^{131}I is 8 days.
         What fraction of the initial ^{131}I concentration remains after 24 days? (2 marks)

b. Use the Maxwell Boltzman distribution curves to explain the effect of:
    i) temperature on the rate of a chemical reaction (3 marks)

    ii) a negative catalyst on the rate of reaction (2 marks)

_____

_____

12.3

a. i) Describe the steps you would take to determine the rate of catalytic decomposition of hydrogen peroxide by titrimetric analysis. (5 marks)

_____

_____

_____

_____

_____

ii) Explain how the results of this experiment can be used to determine the order of the reaction. (3 marks)

_____

_____

_____

_____

Chapter 13

ELECTROCHEMISTRY

During electrochemical processes, (i) Electricity is used to decompose compounds, by a process known as electrolysis. This is carried out in electrolytic cells; (ii) Electricity is produced from chemical reactions carried out in voltaic cells.

Electricity is a flow of charged particles. In metallic conduction, electrons flow through metals, while ions flow in electrolytic conduction through aqueous solutions or molten liquids.

The reactions involved in electrochemical cells are all oxidation-reduction (redox) reactions. In these redox reactions there is competition between atoms and ions to lose or gain electrons.

A voltaic cell is composed of two half cells. Oxidation occurs in one half cell and reduction in the other. Figure 13a below is a diagram of a voltaic cell called the **Daniell cell**.

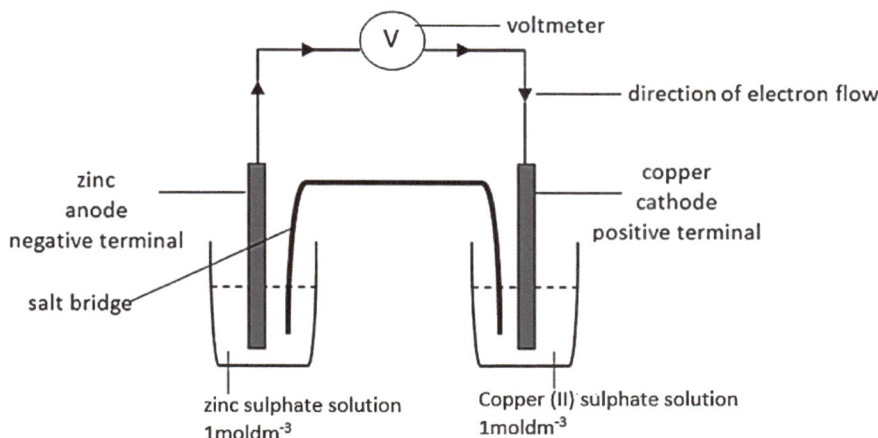

Fig. 13a *diagram of the Daniell Cell*

In the **Daniell cell**, one half cell is the $Zn_{(s)}/Zn^{2+}_{(aq)}$ and the other is the $Cu^{2+}_{(aq)}/Cu_{(s)}$ half-cell. Zn atoms give up electrons more readily than Cu atoms, since zinc is a stronger reducing agent than copper. Zn is therefore oxidised, as each atom loses 2 electrons to form Zn^{2+} ions. These electrons travel through the external conductors, across the voltmeter down the Cu electrode into the Cu half-cell. The Cu^{2+} ions are reduced as each accepts 2 electrons to form $Cu_{(s)}$.

The half-cell reactions are as follows:
$Zn_{(s)}$ - $2e^-$ $\longrightarrow$ $Zn^{2+}_{(aq)}$ (oxidation half cell reaction)
And $Cu^{2+}_{(aq)}$ + $2e^-$ $\longrightarrow$ $Cu_{(S)}$ (reduction half reaction)
Overall redox reaction: $Zn_{(S)}$ + $Cu^{2+}_{(aq)}$ $\longrightarrow$ $Zn^{2+}_{(aq)}$ + $Cu_{(s)}$

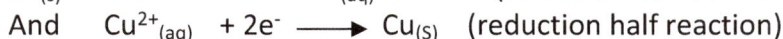

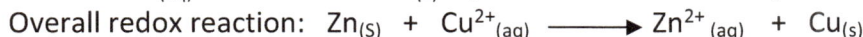

The Cu electrode is the cathode and the positive terminal, while the Zn electrode is the anode and negative terminal of the cell.

The **cathode** is defined as the electrode along which electrons enter an electrochemical cell, and at which reduction occurs.

The **anode** is defined as the electrode along which electrons leave a cell, and at which oxidation occurs.

For an electrochemical cell to function, electrical neutrality must be maintained in both electrolytes. A salt bridge or a porous barrier between the two electrolytes connects the two solutions. A salt bridge is made from a strip of filter paper soaked in a solution of KNO_3 or $NaNO_3$. It maintains the electrical neutrality of the two solutions by allowing excess negative ions from the copper half-cell to flow across to the zinc half-cell where positive Zn^{2+} ions are being produced. By so doing, the electrical neutrality of both solutions is maintained.

The positive ions that were building up in the zinc half-cell are neutralised by the negative ions that have flowed across the salt bridge from the copper half-cell. Ions can flow in both directions across the salt bridge.

KNO_3 and $NaNO_3$ are used to soak the filter paper because all potassium, nitrate and sodium salts are soluble, and therefore ions flowing across it, will not be precipitated.
The salt bridge also completes the electrical circuit, without allowing the solutions to mix.

Standard electrode potentials and the standard hydrogen half-cell

The voltage generated by a cell when two half cells are set up, is called the **electromotive force (e.m.f)**. This is a measure of the potential difference between the two half-cells (electrodes). The unit is in voltage.

The **Standard Electrode Potential** of a half cell $(E^{\varnothing})$ is defined as the electromotive force of the half-cell when set up with the standard hydrogen half-cell at 25°C, 1 atmosphere pressure using 1 moldm^{-3} solutions. The hydrogen half-cell is assigned an electrode potential of zero. Therefore the voltage generated is the standard electrode potential of the half-cell under investigation.

The standard hydrogen half-cell or **standard hydrogen electrode**, consists of hydrogen gas travelling down an inert platinum electrode, at 1 atm pressure and 25° C, bubbling into an electrolyte of 1 moldm^{-3} aqueous H^+ ions.

The electrode is 'platinised' at its base, that is, it is covered with a layer of finely divided platinum. This increases the surface area for the reactions occurring in the half-cell.

Oxidation or reduction can occur at the standard hydrogen electrode, depending on the strength of the half cell under investigation as an oxidising or reducing agent.

When oxidation occurs: $\quad 2H_{(g)} - 2e^- \rightarrow 2H^+_{(aq)}$
When reduction occurs: $\quad 2H^+_{(aq)} + 2e^- \rightarrow H_{2(g)}$

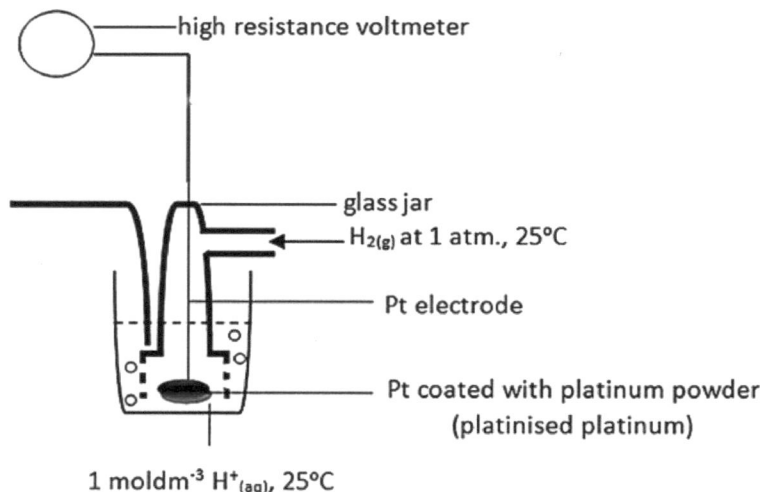

— high resistance voltmeter

— glass jar
— $H_{2(g)}$ at 1 atm., 25°C

— Pt electrode

— Pt coated with platinum powder (platinised platinum)

1 moldm^{-3} $H^+_{(aq)}$, 25°C

Fig.13b *the standard hydrogen electrode/ half cell*

The determination of the standard electrode potential of the zinc half cell

Figure 13c shows the arrangement to determine the standard electrode potential of zinc. Since zinc is a stronger reducing agent than hydrogen, Zn atoms are oxidised to Zn^{2+} ions. The electrons from the Zinc half-cell travel externally through the external conductor, across the voltmeter, down the platinum electrode. On the surface of the platinised platinum, the electrons are accepted by aqueous hydrogen ions to form hydrogen gas. The Zn half-cell is therefore the anode and the negative terminal. The standard hydrogen electrode is the cathode and the positive terminal. Oxidation is occurring in the Zn half-cell and reduction in the hydrogen half-cell.

Since the standard hydrogen electrode is assigned an e.m.f of 0, the reading on the voltmeter represents the standard electrode potential of the zinc half-cell, which is 0.76V.

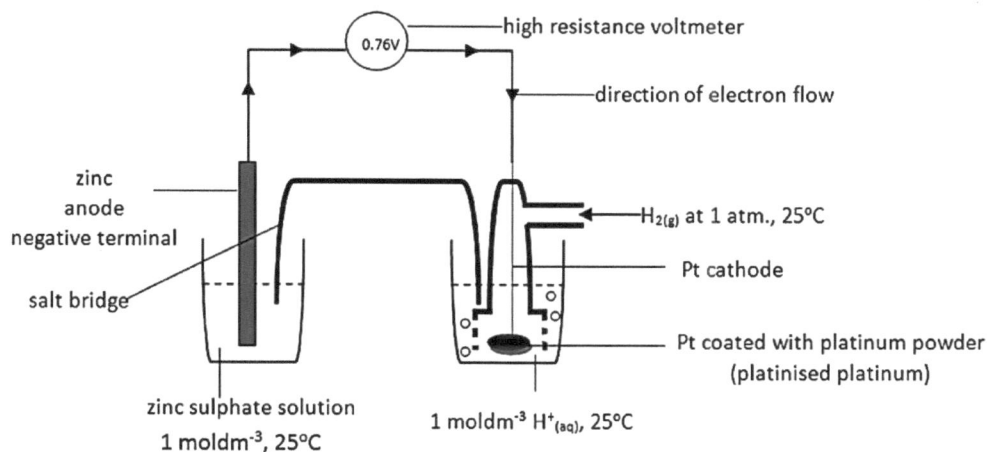

Fig.13c *apparatus for determining the $E^{\varnothing}$ of the zinc half cell*

The determination of the standard electrode potential of the silver half cell

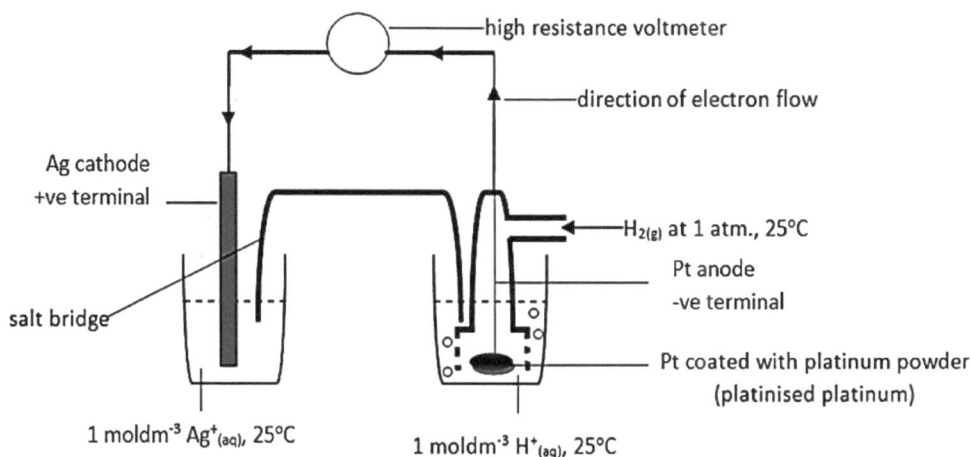

Fig.13d *apparatus for determining the $E^{\varnothing}$ of the silver half cell*

During the determination of the standard electrode potential of silver, oxidation occurs in the hydrogen half-cell and reduction in the silver half-cell. Note that the electrons now flow from the hydrogen half-cell to the silver half cell. The standard hydrogen electrode is the anode and negative terminal, while the silver electrode is the cathode and positive terminal. The reaction occurring at each electrode is as follows:

At the anode: $H_{2(g)} - 2e^- \rightarrow 2H^+_{(aq)}$
At the cathode: $Ag^+_{(aq)} + 1e^- \rightarrow Ag(s)$

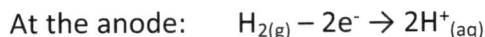

If the standard electrode potential is being determined for a system in which aqueous ions are producing aqueous ions, for example, MnO_4^- /Mn^{2+} or Fe^{2+}/ Fe^{3+}, the two solutions, both 1 moldm^{-3}, are mixed, as shown in the diagram below, and inert platinum is used as the electrode.

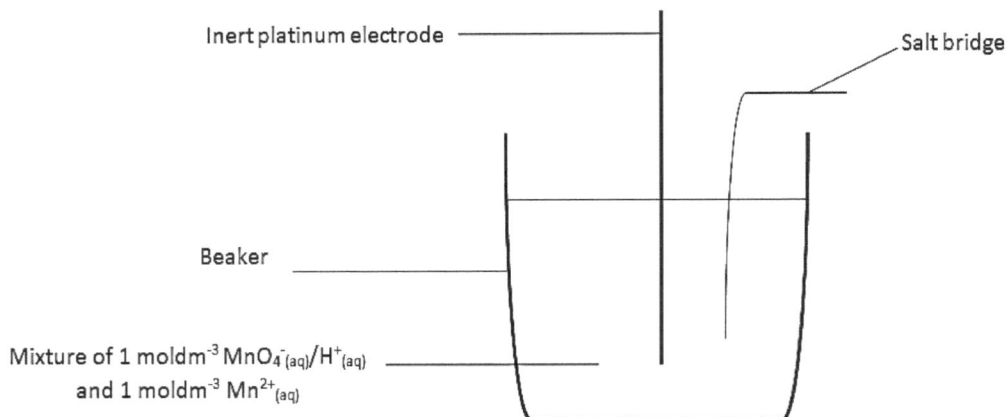

Fig. 13e *half-cell for aqueous reactants and products*

Cell Diagrams or cell statements

A cell diagram shows the combination of the two half-cells and in the order in which the half-cell reactions occur.

For the Daniel Cell, the cell diagram is as follows:

$Zn_{(s)}|Zn^{2+}_{(aq)} \parallel Cu^{2+}_{(aq)} | Cu_{(s)}$ This means that $Zn_{(s)}$ is oxidised to $Zn^{2+}_{(aq)}$ in one half-cell and $Cu^{2+}_{(aq)}$ is reduced to $Cu_{(s)}$ in the other half-cell.

$\parallel$ *represents a salt bridge or a porous barrier between the two half-cells.*

Cell diagrams for the standard electrode potentials of zinc and silver respectively are as follows:

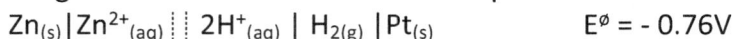

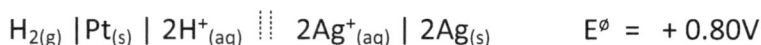

$Zn_{(s)}|Zn^{2+}_{(aq)} \parallel 2H^+_{(aq)} | H_{2(g)} |Pt_{(s)}$ $E^\phi = - 0.76V$

$H_{2(g)} |Pt_{(s)} | 2H^+_{(aq)} \parallel 2Ag^+_{(aq)} | 2Ag_{(s)}$ $E^\phi = + 0.80V$

Interpreting standard electrode potentials
According to the International Union of Pure and Applied Chemistry (IUPAC) Convention:
- Standard half-cell reactions must be written in the form of the reduction.
- If the reduction reaction was spontaneous, then a positive sign (+) is placed in front of the standard electrode potential.
- If the reduction did not occur, then a negative sign (-) is placed in front of the standard electrode potential. (that is, the oxidation reaction was spontaneous)

Therefore:

$Ag^+_{(aq)} + 1e \rightarrow Ag_{(s)}$ $E^\phi = + 0.80V$ since the reduction occurred in this half-cell   (Reduction occurred)

$Zn^{2+}_{(aq)} + 2e \rightarrow Zn_{(s)}$ $E^\phi = -0.76V$ since reduction did not occur in this half-cell. (Oxidation occurred)

Referring to Table 13.1, the negative sign in front of the E⁰ value for the Zn and K half-cells, means that reduction did not occur. Instead oxidation occurred in those two half-cells. This means that both Zn and K are reducing agents, since they were oxidised. Because 2.92V was generated in the K half-cell compared to only 0.76V in the Zn half-cell, it means that K is a stronger reducing agent than Zn.

E⁰	Half-cell reaction
- 2.92	$K^+_{(aq)} + e \rightarrow K_{(s)}$
- 0.76	$Zn^{2+}_{(aq)} + 2e \rightarrow Zn_{(s)}$
+0.34	$Cu^{2+}_{(aq)} + 2e \rightarrow Cu_{(s)}$
+0.80	$Ag^+ + e \rightarrow Ag_{(s)}$
+1.07	$Br_{2(l)} + 2e \rightarrow 2Br^-_{(aq)}$
+1.36	$Cl_{2(l)} + 2e \rightarrow 2Cl^-_{(aq)}$
+1.52	$MnO_4^-{}_{(aq)} + 8H^+_{(aq)} + 5e \rightarrow Mn^{2+}_{(aq)} + 4H_2O_{(l)}$

Table 13.1 *Table of standard electrode potentials*

The positive sign in front of the other E⁰ values means that the reduction was spontaneous; hence, they are all oxidising agents. Therefore, Br_2, Cl_2, and MnO_4^- for example, are all oxidising agents. MnO_4^- is the strongest oxidising agent among them because it has the most positive/highest E⁰ value and K^+ the weakest, having the least positive value.

The table shows that potassium, K is the strongest reducing agent, while the permanganate (vII) ion, MnO_4^-/H^+ is the strongest oxidising agent.

Uses of standard electrode potentials /E⁰ Values

1. To compare the oxidizing and reducing strengths of various oxidizing and reducing agents. (Table 13.1)
2. To calculate the standard cell potential of any combination of half-cells.

The Standard Cell Potential (E⁰$_{cell}$) is defined as the sum of the standard electrode potentials of two half-cells

3. To determine the feasibility of a cell reaction. That is, whether a reaction will occur or not. A cell reaction will occur between two half-cells, if the calculated cell potential is positive and one half-reaction is oxidation and the other is reduction, as opposed to both being oxidation or both being reduction.

Worked example 13.1

a. Use the following half-cell reactions and their E⁰ values

$F_2 + 2e^- \rightarrow 2F^-$ +2.87 V

$Na^+ + 1e \rightarrow Na_{(s)}$ -2.71 V

To predict whether a reaction will occur between:

   i) **F_2 + Na**

Solution

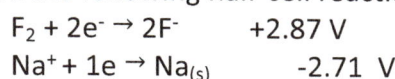

$F_2 + 2e^- \rightarrow 2F^-$	E⁰ = +2.87 V	reduction half
$2Na - 2e \rightarrow 2Na^+_{(s)}$	E⁰ = +2.71 V	oxidation half
$2Na_{(s)} + F_{2(g)} \rightarrow 2F^-_{(aq)} + 2Na^+_{(aq)}$	E⁰$_{cell}$ +5.58V	**This reaction is feasible.**

   ii) **F- + Na+**

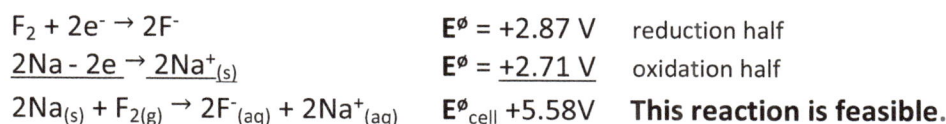

$2F^- - 2e \rightarrow F_2$	E⁰ = -2.87V	
$2Na^+ + 2e \rightarrow 2Na_{(s)}$	E⁰ = -2.71V	
$2F^-_{(aq)} + 2Na^+_{(aq)} \rightarrow 2Na_{(s)} + F_{2(g)}$ E⁰$_{cell}$ −5.58V		**This reaction is not feasible and will not occur.**

   iii) **F_2 + Na+**

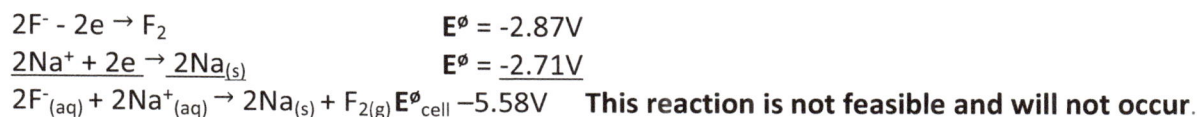

$F_2 + 2e^- \rightarrow 2F^-$ E⁰ = +2.87V

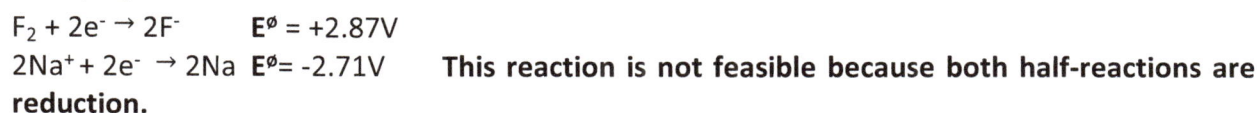

$2Na^+ + 2e^- \rightarrow 2Na$ E⁰= -2.71V   **This reaction is not feasible because both half-reactions are reduction.**

Factors that affect cell potentials

The factors that will affect cell potentials are:
1. Concentration of aqueous ions
2. Temperature
3. Pressure of gases

Concentration of Aqueous Ions

If **the concentration of a product ion is increased**, for example Zn^{2+} in the Daniell cell, according to Le Châtelier's principle, there will be a tendency to decrease its concentration. This occurs as the rate of oxidation of Zn to Zn^{2+} decreases, resulting in a slower rate of electron flow and a decrease in the voltage/ electrode potential of the cell.

If **the concentration of the product ion is decreased,** there will be a tendency to increase its concentration. This occurs as the rate of oxidation of the Zn to Zn^{2+} increases, producing a rapid flow of electrons and a higher voltage or cell potential.

If **the concentration of the reactant ion is increased,** For example, Cu^{2+} ion in the Daniell cell, there will be a tendency, according to Le Châtelier's principle, to decrease its concentration. This occurs as the Cu^{2+} ions are rapidly reduced to $Cu_{(s)}$, creating a high demand and rapid flow for electrons from the anode, resulting in an increase in the electrode potential of the cell.

If the **concentration of the reactant ion is decreased**, there will be a tendency, according to Le Châtelier's Principle, to increase its concentration. This occurs as Cu^{2+} ions are slowly reduced to $Cu_{(s)}$, creating a low demand and slow flow of electrons from the anode. This results in a decrease in the electrode potential of the cell.

Commercial Cells

These are voltaic cells, that are an easy, inexpensive convenient and mobile source of electricity and they generally produce 1.5V of electricity. These cells may be non-rechargeable, called **primary** cells, or rechargeable, called **secondary** cells or accumulators.
A **battery** is a group of cells.

The LeClanché dry cell

This is a Primary cell. It is the commonest, cheapest and most convenient source of electricity. This cell has a short life however and has to be thrown away when it is used up. It is widely used in torches, toys and radios for example.

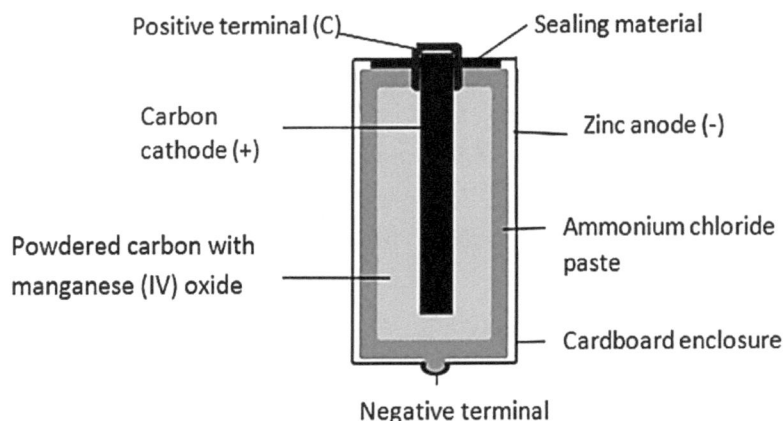

Positive terminal (C)
Sealing material
Carbon cathode (+)
Zinc anode (-)
Powdered carbon with manganese (IV) oxide
Ammonium chloride paste
Cardboard enclosure
Negative terminal

Fig. 13f *A dry cell*

The redox reactions occurring in this dry cell are as follows:

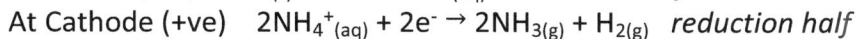

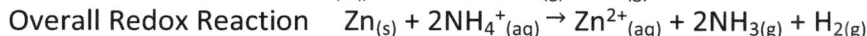

At Anode (-ve) $Zn_{(s)} - 2e^- \rightarrow Zn^{2+}_{(aq)}$ *oxidation half*
At Cathode (+ve) $2NH_4^+_{(aq)} + 2e^- \rightarrow 2NH_{3(g)} + H_{2(g)}$ *reduction half*
Overall Redox Reaction $Zn_{(s)} + 2NH_4^+_{(aq)} \rightarrow Zn^{2+}_{(aq)} + 2NH_{3(g)} + H_{2(g)}$

The gaseous products, $H_{2(g)}$ and $NH_{3(g)}$ produced can coat the surface of the carbon electrode and slow down the reduction reaction occurring. This decreases the efficiency of the cell. The gases are removed by the following reactions:
$H_{2(g)} + 2MnO_{2(s)} \rightarrow Mn_2O_{3(s)} + H_2O_{(l)}$
$NH_{3(g)} + H_2O_{(l)} \rightarrow OH^-_{(aq)} + NH_4^+_{(aq)}$
The NH_4^+ ion is recycled. However, the zinc cannot be replaced.
Therefore, the cell 'dies' when holes appear in the zinc making the cell un-rechargeable.

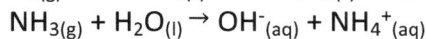

The Lead-Acid Accumulator

This is used in motor vehicles to produce electricity to ignite the fuel as well as to power all electrical parts and devices. A car battery has six 2-volt cells, producing a total of 12V of electricity.

The redox reactions are as follows:

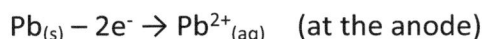

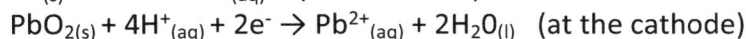

$Pb_{(s)} - 2e^- \rightarrow Pb^{2+}_{(aq)}$ (at the anode)
$PbO_{2(s)} + 4H^+_{(aq)} + 2e^- \rightarrow Pb^{2+}_{(aq)} + 2H_2O_{(l)}$ (at the cathode)
        $\uparrow$
    $(2H_2SO_4)$
$Pb_{(s)} + PbO_{2(s)} + 2H_2SO_{4(aq)} \rightarrow 2PbSO_{4(s)} + 2H_2O_{(l)}$ (overall equation for redox reaction)

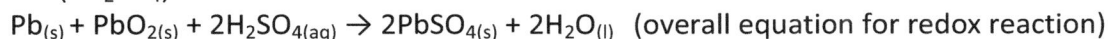

The $PbSO_4$ precipitate can coat the electrode, reducing its efficiency and even causing the battery to 'die'. However, whilst the engine is 'running' and the battery is producing electricity, some of this electricity reverses the reaction above by **electrolysis**, decomposing the $PbSO_4$. This way the reagents are constantly replaced or recharged.

If the $PbSO_{4(s)}$ builds up and becomes **coarse**, for example when a car engine has been off for a long time, the battery can be recharged using electricity from another vehicle using a 'Jump cable'. Over time, however, the fine $PbSO_4$ precipitate will gradually get coarser and builds up. It is now in an inactive non-reversible form and cannot be recharged. The battery therefore 'dies'.

Button cells

These are small and are used in digital watches, hearing aids and calculators for example. They produce about 1.5V and can last for 1 year or more before the voltage drops.

The mercury oxide – zinc cell

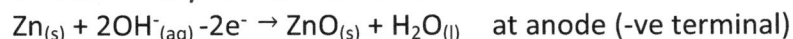

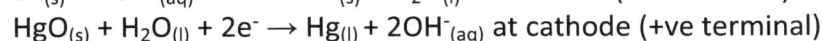

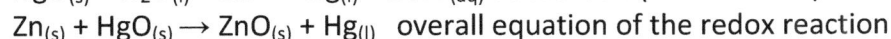

$Zn_{(s)} + 2OH^-_{(aq)} - 2e^- \rightarrow ZnO_{(s)} + H_2O_{(l)}$ at anode (-ve terminal)
$HgO_{(s)} + H_2O_{(l)} + 2e^- \rightarrow Hg_{(l)} + 2OH^-_{(aq)}$ at cathode (+ve terminal)
$Zn_{(s)} + HgO_{(s)} \rightarrow ZnO_{(s)} + Hg_{(l)}$ overall equation of the redox reaction

Other button cells are the **silver oxide - zinc cell, the zinc - air cell** and the **lithium cell**.

The Lithium cell produces 3V of electricity. It is very small in size and more expensive than the others. It is used in heart pacemakers and small electrical items.

The redox reactions are as follows:

$Li_{(s)} - e^- \rightarrow Li^+_{(s)}$ at anode (-ve terminal)

$Li^+_{(s)} + MnO_{2(s)} + e^- \rightarrow LiMnO_{2(s)}$ at cathode (+ve terminal)

$Li_{(s)} + MnO_{2(s)} \rightarrow LiMnO_{2(s)}$ overall equation for the redox reaction

The fuel cell

A fuel cell is a voltaic cell which produces electricity through an electrochemical reaction using fuels such as hydrogen as a reducing agent and oxygen or another, as oxidising agent.

The H_2 - O_2 fuel cell is a primary cell that **never 'dies'** since the fuel, hydrogen and oxygen are constantly replaced.

The electrodes are porous C or porous Ni in contact with an alkaline solution.

The half-cell reactions are as follows:

$2H_{2(g)} + 4OH^-_{(aq)} - 4e^- \rightarrow 4H_2O_{(l)}$ at anode (-ve terminal)

$O_{2(g)} + 2H_2O_{(l)} + 4e^- \rightarrow 4OH^-_{(aq)}$ at cathode (+ve terminal)

$2H_{2(g)} + O_{2(g)} \rightarrow 2H_2O_{(l)}$ overall equation for the redox reaction

The gases are circulated under pressure to come in contact with the electrodes.
They are used to power cars and the Apollo moon vehicles.

Disadvantages – Expensive to produce, can only store a small amount of energy, fossil fuels are still needed to produce hydrogen, hydrogen is flammable and the cell could explode.large and heavy.

Advantages – No greenhouse gas emissions, inexpensive maintenance as there are no electrodes to replace, and the fuels O_2 and H_2 are fed in continuously. Pollutant free, pure H_2O is the only by-product.

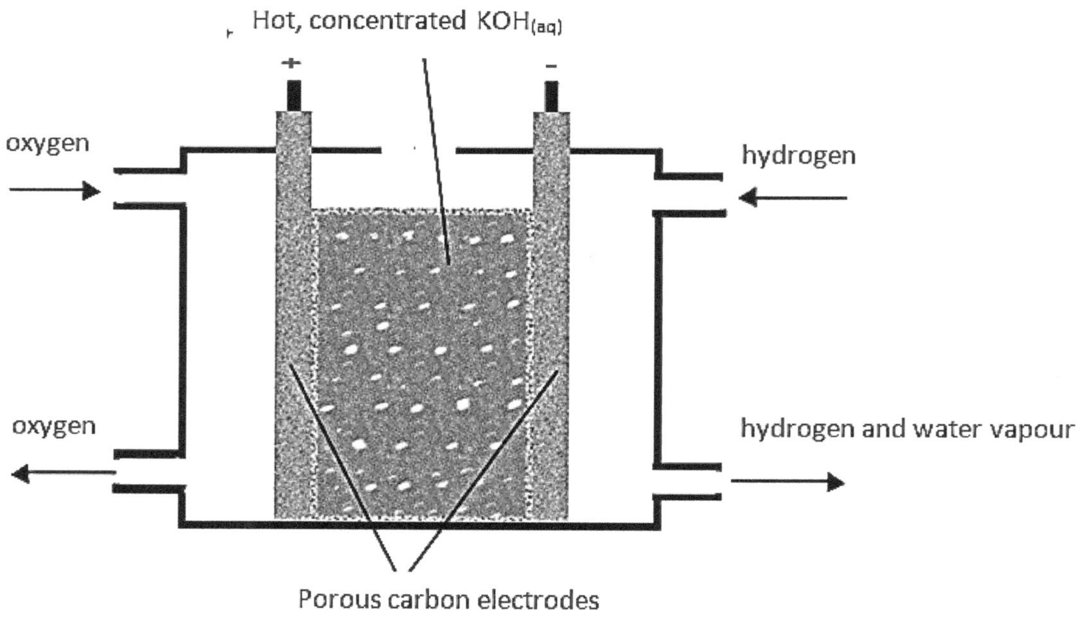

Fig. 13g *A simple fuel cell*

Electrical cars

In recent times battery powered electrical cars are being developed by many car manufacturers. These vehicles are small in size and develop a moderate speed. They are powered by one or more electric motors, using energy stored in rechargeable batteries.These cars are presently available in the Caribbean, United States, Europe and other countries. They

are ecologically friendly producing no pollutants. The batteries are recharged at charging stations installed in homes or public places.
Presently the battery range is between 280 km to 350 km. 750,000 electric cars were registered in 2016.

Exercise 13.1

a. i) Use E^θ values from a data booklet to determine if zinc metal will displace copper from an aqueous solution of copper (II) ions. Explain your answer. (3 marks)

_____

_____

_____

   ii) Write a balanced equation for the reaction above and describe any changes observed. (3 marks)

_____

_____

_____

b. i) Draw a fully labelled diagram of the electrochemical cell for the following two half-cells under standard conditions: $Fe^{2+}_{(aq)} \rightarrow Fe^{3+}_{(aq)}$ and $MnO_4^-{}_{(aq)} \rightarrow Mn^{2+}_{(aq)}$ in acid pH (5 marks)

   ii) Write the fully balanced equation for the above reaction. (1 mark)

_____

   iii) Explain how the electrode potential of the above cell would change if the concentration of MnO_4^-(aq) is increased. (2 marks)

_____

_____

13.2

a. Define the terms:

   i) Standard electrode potential (1 mark)

_____

_____

   ii) Standard cell potential (1 mark)

_____

_____

b. A simple rechargeable cell may be constructed by dipping two lead electrodes into aqueous lead (II) nitrate and passing a current for a few minutes. During this process, lead (IV) oxide is deposited on one of the electrodes and the other electrode reacts. When the power source is disconnected and a bulb is connected across the two electrodes, the bulb lights for a time as the cell discharges.

   i) Choose two half-equations to construct the full equation for the reaction that occurs during this discharge. And calculate the $E^\varnothing$ value for this cell reaction. (3 marks)

   ii) In the lead-acid accumulator, similar reactions occur but the electrolyte of $H_2SO_{4(aq)}$ causes Pb^{2+} ions to be precipitated as $PbSO_{4(s)}$. The e.m.f of this cell is 2.0V. Explain the difference between this e.m.f and the $E^\varnothing_{(cell)}$ value calculated in (a) above. (2 marks)

_____

_____

Chapter 14

Chemistry of the elements and their compounds

Structure of the Period Table

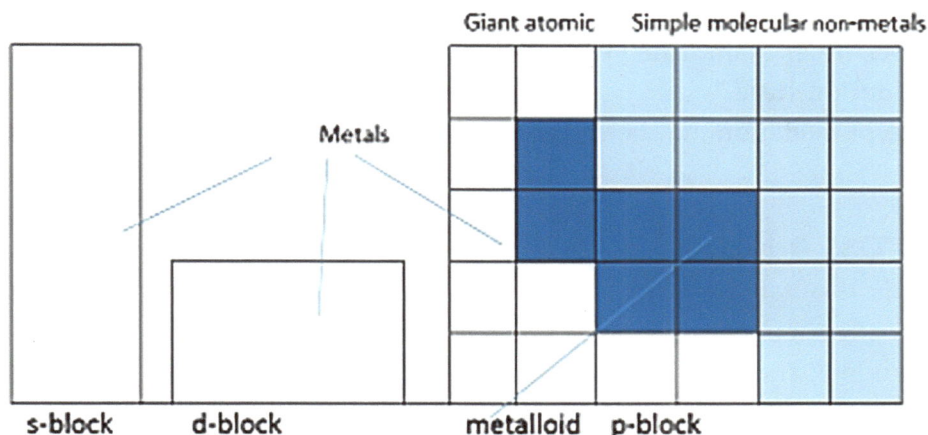

Fig. 14a *Structure of the Periodic Table*

Most of the elements in the Periodic Table are metals. The non-metals consist mostly of simple molecules, some, notably carbon and silicon have giant atomic structures. It is the structure and the bonding within them, which are largely responsible for the variations in physical properties. In **Group I**, all the elements from lithium to caesium have body-centred cubic giant metallic structures. Francium is extremely radioactive, so is very difficult to investigate.

In Group 0/Group 8, all the elements from neon to radon are monatomic gases, consisting of single atoms. For all members of **Group 7**, from fluorine to astatine, the structure is diatomic molecules, X_2, whether the element exists as a gas F_2 and Cl_2, liquid Br_2, or solid I_2 and At_2 at room temperature.

Across a Period, the general trend is from giant metal structures to simple molecular non-metals. **Group 2** begins with an element, beryllium, which is a good conductor of electricity but has few of the chemical properties of a metal. **Group 3** begins with a non-metal, boron, which has a very high melting point and a giant atomic structure. But the next member in each case is a metal, magnesium and aluminium respectively.

In **Group 4**, carbon is a non-metal. Silicon is also a non-metal, though with some metallic properties; the same is true of germanium. These are best described as **metalloids**. Tin and lead are metals.

Chemical Periodicity

Approximately 70 of the first 90 elements in the Periodic Table are metals. Metals are on the left of the Table, the most reactive of them at the bottom left. Non-metals are on the right, with the most reactive at the top right excluding the noble/inert gases.

Typical metals are good conductors of heat and electricity, easily lose electrons to form positive ions, form basic oxides, and form ionic solid chlorides.

Typical non-metals are non-conductors, easily gain electrons to form negative ions, and form acidic or neutral oxides and volatile, often liquid covalent chlorides.

- Metals typically form compounds by ionic bonding with non-metals.
- They can also form alloys when mixed with each other.
- Non-metals form either ionic bonds with metals or covalent bonds with each other.
- The properties of some elements are neither completely metallic nor non-metallic, and the bonding in many compounds is neither clearly ionic nor covalent.
- Elements, other than those in Groups 1 and 2, are often capable of existing in more than one **oxidation state**.

Oxidation state or oxidation number gives the combining power of an atom in a compound or ion.

Trends across a Period

A Period is a horizontal row that starts with an extremely reactive metal in Group 1. The elements gradually become more non-metallic across the Period with the greatest non-metallic character in Group 7. The period ends with a chemically inert/noble gas. Metals in general have low first ionization energy and metallic character decreases across a Period. Non-metallic character increases across a Period.

Oxidation State

The maximum oxidation state of an element is equal to the Group number.

Group	1	2	3	4	5	6	7	0/8
Maximum oxidation state								
Period Li → Ne	+1	+2	+3	+4	+5	-2	-1	0
Period Na → Ar	+1	+2	+3	+4	+5	+6	+7	0
Common oxidation states	+1	+2	+3	+4	+5	+6	+7	0
				+2	+3	+4	+5	
					-3	-2	-1	

Table 14.1 *Oxidation states of some elements*

Trends down a Group

Groups are vertical columns in the Periodic Table. Down a Group there is an:

- Increase in atomic number
- Increase in atomic radius
- Increase in ionic radius
- Increase in electropositive character
- Increase in metallic character
- Decrease in first ionization energy

Electropositivity is the tendency for atoms to give up electrons and form positively charged ions. In the Periodic Table, down any Group there is an increase in metallic character and electropositivity. This trend can be explained by the increasing shielding and atomic radius down the group. The most reactive metal is therefore found at the bottom left of the Table.

The most reactive non-metal is found at the top right of the Table, having the highest electronegativity. (group 0/8 of inert gases excluded).

Electronegativity is the tendency of an atom in a covalent bond, to attract the bonding pair of electrons towards itself. Fluorine is the most electronegative atom with caesium and francium the least electronegative. Electronegative atoms gain electrons and form negatively charged ions during ionic bonding. Electronegativity increases as the atomic radius decreases.

In some Groups the first element exhibits anomalous behaviour. In these instances, the difference in properties between the first and second member of the Group is often greater than that between the other members of the Group altogether. This is because of the small atomic radius and consequent high electronegativity of the first element, for example lithium, beryllium and boron.

The Inert Pair Effect

In a Group where the elements can exist in more than one oxidation state, for example, Group 4, as the atomic number increases down the Group, the lower oxidation state tends to become more stable. In Group 4, the +4 oxidation state is more stable in C, as in CO_2, while the +2 oxidation state is more stable in Sn and Pb as in SnO and PbO respectively. This is explained as follows:

As the group is descended the atomic radius increases.

At the top of the group carbon shares all 4 valence electrons in covalent bonding.

As shielding increases, and the atomic radius increases down the Group, the two, p electrons in the valence shell ($s^2 p^2$) are less strongly attracted to the nucleus and are lost easily, forming ions with +2 oxidation state, for example, Pb^{2+}. The 2, s electrons are therefore not involved in bonding and are referred to as an inert pair.

Physical properties across Period 3

Period 3 elements	Na	Mg	Al	Si	P	S	Cl	Ar
Bonding	metallic			covalent →→				
Structure	giant metallic			giant atomic	←simple molecular →→			
MP ^{0}C	98	651	660	1410	44	114	-101	-189
BP ^{0}C	890	1120	2450	2680	280	445	-34	-186
Density/25^{0}C/g cm$^-$$_3$	0.997	1.74	2.70	2.33	1.82	2.07	1.57	1.40 (in liquid state at BP)
Atomic radius/nm	0.186	0.160	0.143	0.117	0.110	0.104	0.099	0.192

Table 14.2 *some trends across Period 3*

The explanation for the **atomic radius of argon being larger** than that of the other elements across the period is as follows: There are strong forces of attraction between atoms of the other elements in the period, due to metallic and covalent bonding. The distance between two atomic nuclei is therefore less than that between two argon unbonded argon atoms that are merely 'touching' due to weak Vander Waal forces. The radii are therefore referred to as the **metallic radius, covalent radius and Vander Waal radius** respectively.

Trends in reactivity of Period 3 elements with water, oxygen and chlorine

Reagent	Sodium	Magnesium	Aluminium	Silicon	Phosphorus P_4	Sulphur S_8	Chlorine Cl_2	Argon
Water	Violent reaction with cold water. Strongly alkaline solution	Reacts with hot water or steam. MgO is slightly soluble in water; alkaline solution	Thin, tough surface layer of oxide prevents reaction	No reaction	No reaction	No reaction	React with water to form an acidic solution which bleaches	No reaction
Oxygen	Burns when ignited	Burns with intense white light when ignited	Burns readily when ignited in oxygen	Burns readily when ignited in oxygen	Can ignite in air. Burns fiercely in oxygen when ignited	Burns when ignited in oxygen	No reaction	No reaction
Chlorine	Burns when ignited	Burns when ignited	Heat in dry chlorine and product sublimes off	Heat in dry chlorine	Warm in dry chlorine	Pass dry chlorine over molten sulphur		No reaction

Table 14.3 *The reactions of the elements of Period 3 with water, oxygen and chlorine*

Some equations for the above reactions:

(i) Water

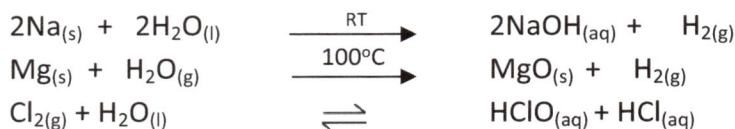

$$2Na_{(s)} + 2H_2O_{(l)} \xrightarrow{RT} 2NaOH_{(aq)} + H_{2(g)}$$
$$Mg_{(s)} + H_2O_{(g)} \xrightarrow{100°C} MgO_{(s)} + H_{2(g)}$$
$$Cl_{2(g)} + H_2O_{(l)} \rightleftharpoons HClO_{(aq)} + HCl_{(aq)}$$

(ii) Oxygen

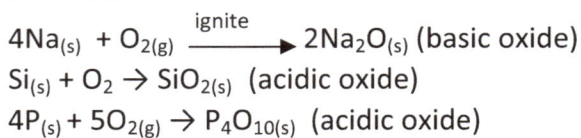

$$4Na_{(s)} + O_{2(g)} \xrightarrow{ignite} 2Na_2O_{(s)} \text{ (basic oxide)}$$
$$Si_{(s)} + O_2 \rightarrow SiO_{2(s)} \text{ (acidic oxide)}$$
$$4P_{(s)} + 5O_{2(g)} \rightarrow P_4O_{10(s)} \text{ (acidic oxide)}$$

(iii) Chlorine

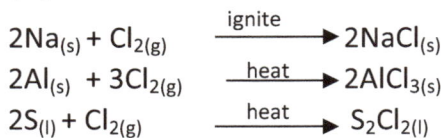

$$2Na_{(s)} + Cl_{2(g)} \xrightarrow{ignite} 2NaCl_{(s)}$$
$$2Al_{(s)} + 3Cl_{2(g)} \xrightarrow{heat} 2AlCl_{3(s)}$$
$$2S_{(l)} + Cl_{2(g)} \xrightarrow{heat} S_2Cl_{2(l)}$$

The oxides of the elements of Period 3

Period 3 produces a mixture of basic and acidic oxides. A **basic oxide or hydroxide** is one which reacts with an acid to form a salt and water only.

Metal oxides are generally basic. If a metal oxide is soluble in and reacts with water the solution is called an **alkali**.

$$Na_2O_{(s)} + H_2O_{(l)} \xrightarrow{RT} 2NaOH_{(aq)}$$

$$Na_2O_{(s)} + 2HCl_{(aq)} \xrightarrow{RT} 2NaCl_{(aq)} + H_2O_{(l)}$$

An **acidic oxide** is one which reacts with an alkali or insoluble basic metal oxide to form a salt and water.

Oxides of non-metals are generally acidic. Acidic oxides dissolve in water to form an **acidic solution**;

$$SO_{3(g)} + H_2O_{(l)} \longrightarrow H_2SO_{4(aq)}$$

Silicon oxide, SiO_2 is insoluble in water, but is still acidic since it reacts with an alkali to form a salt and water.

$$SiO_{2(s)} + 2NaOH_{(aq)} \xrightarrow[\substack{concentrated \\ solution}]{heat} Na_2SiO_{3(aq)} + H_2O_{(l)}$$

An **amphoteric oxide or hydroxide** is one which can react with both acids and bases, for example, Aluminium oxide, Al_2O_3.

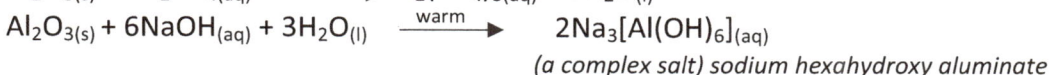

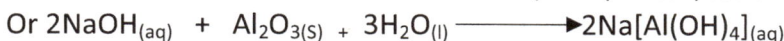

$$Al_2O_{3(s)} + 3H_2SO_{4(aq)} \xrightarrow{warm} Al_2(SO_4)_{3(aq)} + 3H_2O_{(l)}$$

$$Al_2O_{3(s)} + 6NaOH_{(aq)} + 3H_2O_{(l)} \xrightarrow{warm} 2Na_3[Al(OH)_6]_{(aq)}$$

(a complex salt) sodium hexahydroxy aluminate

Or $2NaOH_{(aq)} + Al_2O_{3(S)} + 3H_2O_{(l)} \longrightarrow 2Na[Al(OH)_4]_{(aq)}$

Neutral oxides react with neither acid nor base. Examples are NO, N_2O and CO. Although water is a neutral oxide, it can act as both an acid and a base, donating a proton and accepting a proton, respectively. The oxides change across Period 3, from strongly basic through to amphoteric to strongly acidic. This correlates well with the change in the elements from electropositive on the left to electronegative on the right, and the change in the structures of their oxides from giant ionic to simple molecular.

The table below summarizes the properties, bonding and structure of the oxides of Period 3.

	Sodium	Magnesium	Aluminium	Silicon	Phosphorus P_4	Sulphur S_8	Chlorine
Formula Other oxides	Na_2O Na_2O_2	MgO	Al_2O_3	SiO_2	P_4O_{10} P_4O_6	SO_3 SO_2	Cl_2O_7 Cl_2O
Physical state at RT	Solid	Solid	solid	solid	solid	gas	liquid
MP/°C	1275	2852	2027	1610	Sublimes at 300	-17	-92
BP/°C	Sublimes at melting point	3600	2980	2230		-45	80
Bonding	Ionic	Ionic	Ionic and covalent	covalent	covalent	covalent	covalent
Structure	Giant ionic	Giant ionic	Giant ionic	Giant molecular	Simple molecular	Simple molecular	Simple molecular
Electrical conductivity when molten	Good	Good	Good	Semi-conductor	None	None	None
Reaction with water	Dissolves. Forms alkaline solution, $NaOH_{(aq)}$.	Only slightly soluble. Forms $Mg(OH)_2$.	Does not react.	Does not react.	P_4O_{10} Reacts to form H_3PO_4,	SO_2 forms H_2SO_3. SO_3 forms H_2SO_4.	Cl_2O_7 Forms the acid $HClO_4$.
Nature of oxide	Basic	Basic	Amphoteric	Acidic	Acidic	Acidic	Acidic
Equations	$Na_2O_{(s)} + H_2O_{(l)}$ → $2 NaOH_{(aq)}$	$MgO_{(s)} + H_2O_{(l)}$ → $Mg(OH)_{2(s)}$			$P_4O_{10(s)} + 6H_2O_{(l)}$ → $4H_3PO_{4(aq)}$	$SO_{3(g)} + H_2O_{(l)}$ → $H_2SO_{4(aq)}$	$Cl_2O_{7(l)} + H_2O_{(l)}$ → $2HClO_{4(aq)}$

Table 14.4 *the oxides of the elements of Period 3*

The chlorides of the elements of Period 3

For the chlorides of Period 3 there is a change from ionic to covalent across the Period as the elements become more electronegative. The chlorides dissolve in water. Sodium chloride dissolves to form a neutral solution of pH 7.

$$NaCl_{(s)} \xrightarrow[\text{RT}]{H_2O} Na^+_{(aq)} + Cl^-_{(aq)}$$

Aluminium chloride is volatile and reacts violently with water which is typical covalent behaviour, to form an acidic solution of low pH between 2 to 3.

$$2AlCl_{3(s)} + 6H_2O_{(l)} \longrightarrow 2Al(OH)_{3(s)} + 6HCl_{(aq)}$$

The behaviour of aluminium chloride in water is explained in terms of the Al^{3+} ion, having vacant d orbitals as well as a high charge and small size (high charge density). Due to the high polarising power of the Al^{3+} ion therefore, it attracts and bonds to the $O^{\delta-}$ atom in H_2O, as a lone pair on the oxygen atom overlaps with a vacant d orbital in Al^{3+}. This weakens the O-H bonds in the water molecule, causing them to break, releasing aqueous $H^+_{(aq)}$ and $Cl^-_{(aq)}$ ions. The aluminium chloride acts as a Lewis acid, accepting an electron pair. Aluminium oxide or aluminium hydroxide is also formed. This process is known as **hydrolysis**.

The following, Table 14.5 summarizes the properties, bonding and structure of the chlorides of Period 3.

	Sodium	Magnesium	Aluminium	Silicon	Phosphorus P_4	Sulphur S_8	Chlorine	Argon
Formula	NaCl	$MgCl_2$	$AlCl_3$	$SiCl_4$	PCl_3 PCl_5	S_2Cl_2 SCl_2, SCl_4	Cl_2	None formed
Physical state at RT	Solid	Solid	Solid	Liquid	Liquid	Liquid	Gas	
M.P./°C	801	714	178	-70	-112	-80	-101	
B.P./°C	1413	1412	Sublimes to Al_2Cl_6	58	76	136	-35	
Bonding	Ionic	Ionic	covalent	covalent	covalent	covalent	covalent	
Structure	Giant ionic	Giant ionic	Simple molecular	Simple molecular	Simple molecular	Simple molecular	Simple molecular	
Electrical conductivity when molten	Good	Good	Poor	None	None	None	None	
Reaction with water	Dissolves easily	Dissolves easily	Fumes of HCl produced	Fumes of HCl produced	Fumes of HCl produced	Fumes of HCl produced	Some reaction with water	
pH of solution	Neutral 7	Weakly acidic / 5-6	Acidic 2 - 3	Acidic 2 - 3	Acidic 2 - 3	Acidic 2 - 3	Acidic 2 - 3	
Equations	$NaCl_{(s)}$+aq $\rightarrow Na^+_{(aq)}$ $+ Cl^-_{(aq)}$	$MgCl_2$ + aq $\rightarrow Mg^{2+}_{(aq)}$ + $2Cl^-_{(aq)}$	$2AlCl_{3(s)}$ + $6H_2O_{(l)} \rightarrow$ $2Al(OH)_{3(s)}$ $+ 6HCl_{(aq)}$	$SiCl_{4(l)}$ + $4H_2O_{(l)} \rightarrow$ $Si(OH)_{4(s)}$ $+ 4HCl_{(aq)}$	$PCl_{3(l)}$ + $3H_2O_{(l)} \rightarrow$ $H_3PO_{3(aq)}$ + $3HCl_{(aq)}$ $PCl_{5(s)}$ + $5H_2O_{(l)} \rightarrow$ $H_3PO_{4(aq)}$ + $5HCl_{(aq)}$	Various other products for example, H_2S, SO_2, H_2SO_3, H_2SO_4.	$Cl_{2(g)}$ $+H_2O_{(l)}$ $\rightleftharpoons$ $HClO_{(aq)}$ + $HCl_{(aq)}$	

Table 14.5 *The chlorides of the elements of Period 3*

Uses of elements and their compounds of Period 3

1. **Aluminium** is widely used to make pots and pans, aircraft bodies, windows and door frames.

2. **Aluminium hydroxide** is used in antacid medications to neutralise stomach acid produced by acid reflux.

3. **White phosphorus** is used in flares.

4. **Red phosphorus** is used in matches and at the side of match boxes to start a flame for igniting fuels for cooking, for example.

5. **Argon** is used in fluorescent and incandescent lighting.

Exercise 14.1
1. Describe and explain the trend in atomic radius across Period 3. (3 marks)

_____

_____

_____

_____

2. Aluminium oxide is amphoteric. Explain the meaning of this term. Include appropriate equations in your answer. (3 marks)

_____

_____

_____

_____

3. i) State the bonding in aluminium chloride. (1 mark)

_____

ii) Explain why this type of bonding is found in aluminium chloride. (2 marks)

_____

_____

_____

4. i) Magnesium chloride forms a weakly acidic solution in water. Sodium chloride forms a neutral solution whereas aluminium chloride forms a strongly acidic solution. Explain these three observations. (3 marks)

ii) Write balanced equations for the reaction between aluminium chloride and water and between sodium chloride and water. (2 marks)

Chapter 15

Introduction

The elements of Group 2 are often called the **alkaline earth metals**. They are as follows:

Beryllium	Be	$[He]2s^2$
Magnesium	Mg	$[Ne]3s^2$
Calcium	Ca	$[Ar]4s^2$
Strontium	Sr	$[Kr]5s^2$
Barium	Ba	$[Xe]6s^2$
Radium	Ra	$[Rn]\ 7s^2$

Properties of the elements of Group 2

Beryllium has anomalous properties because of its small size and will not be considered. Radium is also not considered as all its isotopes are radioactive.

The alkaline earth metals from magnesium to barium are white metals with **giant metallic structures**. They have low melting and boiling points compared to transition metals like iron. They are good conductors of heat and electricity, and they burn in air with characteristic flame colours. Magnesium burns with a white flame, calcium brick-red, strontium red and barium green.

The metals of Group 2 all have two electrons in their valence shell. These are lost when they form an ion of oxidation number +2 in their compounds, such as Mg^{2+} and Ca^{2+}. They are less reactive than Group 1 metals, because they must lose two valence electrons, whereas Group 1 metals lose only one.

The table below summarizes some physical properties of Group 2 elements.

Element	Mg	Ca	Sr	Ba
Atomic number	12	20	38	56
Metallic/ atomic radius/nm	0.160	0.197	0.215	0.224
Ionic radius/nm	0.072	0.100	0.113	0.136
First ionisation energy/kJmol^{-1}	738	590	550	503
Second ionisation energy/kJmol^{-1}	1451	1145	1064	965
Third ionisation energy/kJmol^{-1}	7733	4912	4210	3,600
Melting point/°C	649	839	769	725
Boiling point/°C	1107	1484	1384	1640

Table 15.1 Physical *properties of Group II elements*

As the group is descended, the atomic radius increases and the first ionisation energy decreases. Reactivity therefore increases down the group.

With increasing atomic radius down the group, the metallic bond gets weaker and melting point decreases.
The density of the metals increases down the group as the relative atomic mass increases.

The **standard electrode potentials** of the alkaline earth metals ($E^{\emptyset}_{M^{2+}/M}$), range from -2.37V to -2.90V (Table 15.2). This shows that they are strong reducing agents, and this property dominates their chemistry.

Element	$E^{\emptyset}_{M^{2+}/M}$ /V		
magnesium	-2.37		Increasing reducing properties down the Group.
calcium	-2.87		
strontium	-2.89		
barium	-2.90		

Table 15.2 *Standard electrode potentials of Group II elements in contact with solutions of their ions*

Summary of general properties

The general properties of the Group 2 elements, magnesium to barium are as follows:
- They are all metals with strong metallic bonds
- They are good conductors of heat and electricity.
- Their compounds are all white solids and colourless solutions.
- In all their compounds they have an oxidation number of +2.
- Their compounds are ionic.
- Their oxides and hydroxides are basic.
- They reduce acids to hydrogen gas.
  Compared with the metals of Group I:
- They are harder and denser.
- They have higher melting points.
- They exhibit stronger metallic bonding because they have two valence electrons instead of one.

Uses of some Group 2 elements and their compounds

The elements of Group 2 and their compounds are widely used in various ways in industry and commerce.

- **Magnesium** is used in flares, incendiary bombs and tracer bullets and as a sacrificial anode on steel objects and bridges.
- **Magnesium hydroxide** is used in indigestion remedies to neutralise stomach acid.
- **Magnesium oxide** is used for the lining of furnaces.
- **Calcium carbonate** is used in making cement.
    Common names of calcium carbonate are **limestone**, **marble** and **chalk**.
- **Calcium hydroxide solid (slaked lime)** is used in lime mortars in the construction industry, and in agriculture to reduce the acidity of the soil.
- **Calcium oxide (quick lime)** is used in cement, mortar and plaster manufacture as well as in agriculture to reduce the acidity of the soil. It also has a role in purifying iron.
- **A solution of calcium hydroxide (lime water)** is used as a test for carbon dioxide gas.
- **Plaster of Paris** is used to set broken bones and for moulds is an insoluble form of calcium sulphate, ($CaSO_4.H_2O$).

- **Barium sulphate in suspension** is given to patients as a 'barium meal' and makes any imperfections in the alimentary canal visible by X-ray photography. Diseases in the digestive tract can therefore be diagnosed.

Reaction of Group 2 metals with water

All these metals reduce water to hydrogen. The reaction with magnesium occurs with steam:

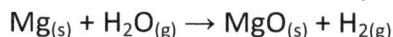

$$Mg_{(s)} + H_2O_{(g)} \rightarrow MgO_{(s)} + H_{2(g)}$$

The other metals react readily with cold water and form white precipitates of the hydroxide.

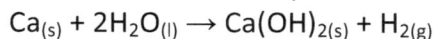

$$Ca_{(s)} + 2H_2O_{(l)} \rightarrow Ca(OH)_{2(s)} + H_{2(g)}$$

The reaction with dilute acids is similar. The metals displace hydrogen as they reduce dilute acids, being stronger reducing agents than hydrogen.

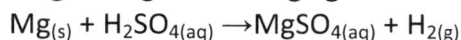

$$Mg_{(s)} + H_2SO_{4(aq)} \rightarrow MgSO_{4(aq)} + H_{2(g)}$$

The reaction of Group 2 metals with oxygen

The reactions, once started, are vigorous and the oxide is formed.

$$2Mg_{(s)} + O_{2(g)} \rightarrow 2MgO_{(s)}$$

The Group 2 metal oxides are all white solids, and they are normally prepared by heating the carbonates, nitrates or hydroxides. The high charge on the metal cation results in strong ionic bond and a high lattice enthalpy, and therefore high melting point of the oxides.

$$CaCO_{3(s)} \xrightarrow{heat} CaO_{(s)} + CO_{2(g)}$$

The reactions of the oxides of Group 2 with water

The general reaction of the metal oxides with water is:

$$MO_{(s)} + H_2O_{(l)} \rightarrow M(OH)_{2(aq)}$$

Both magnesium oxide and magnesium hydroxide are only sparingly soluble in water. Magnesium hydroxide forms a white suspension in water called milk of magnesia that is useful as a mild alkali, to relieve indigestion.

When water is added to calcium oxide it swells and steams, and eventually disintegrates into a white powder, which is calcium hydroxide. This reaction is highly exothermic.

$$CaO_{(s)} + H_2O_{(l)} \rightarrow Ca(OH)_{2(s)} \quad \Delta H \text{ –ve}$$

The stability of the carbonates and nitrates to heat

The Group 2 metals form metal carbonates of formula XCO_3 and metal nitrates of formula $X(NO_3)_2$. When heated, the carbonates decompose to form the metal oxide and carbon dioxide gas:

$$MgCO_{3(s)} \rightarrow MgO_{(s)} + CO_{2(g)}$$

When heated, the nitrates decompose to form the metal oxide, nitrogen dioxide gas and oxygen gas:

$$2Ca(NO_3)_{2(s)} \rightarrow 2CaO_{(s)} + 4NO_{2(g)} + O_{2(g)}$$

Both the carbonates and the nitrates decompose at higher temperatures as the group is descended. This trend relates to the size of the metal ion and the stability of the metal oxide formed as explained below.

1. The size of the metal ion and its polarising power.

Small ions have greater polarising power than large ions, because of their high charge density.

Both the carbonate and the nitrate anions are large and this makes them easily polarisable. Therefore, the small magnesium ion can polarise the large anion, attracting the electrons towards itself. This distorts the shape of the anion and encourages decomposition. The large barium ion, however, has little tendency to polarise the anion, and decomposition is not favoured. The other metal ions fall between these two extremes, and so the decomposition temperatures rise as the Group is descended.

2. The stability of the metal oxide formed.

Fig. 15a *factors influencing the decomposition of the carbonates of Group 2 elements*

When the carbonates and nitrates decompose the metal oxide is formed.

The lattice energy released, when magnesium oxide is formed, is high. This is due to the small size and high charge of the Mg^{2+} and O^{2-} ions. MgO is therefore stable and its formation is favoured.

The lattice energy of barium oxide, however, is small due to the large sized Ba^{2+} ion. BaO is therefore less stable, and is not readily formed. The stability of the oxides formed decreases down the Group, and the decomposition of the carbonate and nitrate becomes less favoured as the Group is descended.

The solubility of the Group 2 sulphates

The sulphates of the Group 2 metals become less soluble as the group is descended.

This trend is explained by considering the lattice energy and the hydration energy, which are the energy changes involved when an ionic solid dissolves in water. The first stage of the process is the separation of the ions as follows:

$$M^{2+}SO_4^{2-}{}_{(s)} \longrightarrow M^{2+}{}_{(g)} + SO_4^{2-}{}_{(g)} \quad \Delta H \text{ +ve}$$

This is an endothermic reaction, requiring the reverse lattice energy. As the ionic radius increases down the Group, the lattice energy required gets progressively lower.

The second stage is the hydration of the separate gaseous ions, when they are surrounded by water molecules. This is an exothermic reaction, releasing energy. As the size of the cation increases on descending the Group, this enthalpy change also becomes less exothermic. Small, highly charged ions have large enthalpy changes of hydration because they can attract and form stronger bonds with the water molecules.

$$M^{2+}{}_{(g)} + SO_4^{2-}{}_{(g)} + aq \longrightarrow M^{2+}{}_{(aq)} + SO_4^{2-}{}_{(aq)} \quad \Delta H \text{ -ve}$$

At the top of the Group, the endothermic enthalpy change of lattice separation is large. However the exothermic enthalpy change of hydration is also large and more than compensates for this. The sulphate therefore has a high solubility. At the bottom of the Group, although the endothermic enthalpy change of lattice separation is lower, the enthalpy change of hydration is even lower, and is not exothermic enough to give such a high solubility. Solubility is therefore low. See Table 15.3 below.

Element	Solubility of the sulphate (mole per 100g of water)
Magnesium	1.83×10^{-1}
Calcium	1.1×10^{-3}
Strontium	7.1×10^{-5}
Barium	9.0×10^{-7}

Table 15.3 *The solubility of the sulphates of Group 2 elements*

The solubility of the hydroxides of Group 2

The solubility of the hydroxides increases down the Group, unlike the sulphates.

This is explained in terms of the smaller size of the hydroxide, OH⁻ anion compared to the large sulphate SO_4^{2-} anion. The large hydration energy released when the relatively small OH⁻ is hydrated, is capable of compensating for the reverse lattice energy, which is decreasing as the Group is descended.

Exercise 15.1

a. Including relevant equations, describe the reactions of the Group 2 metals, magnesium and barium with

i) oxygen (2 marks)

ii) water (2 marks)

Write equations where appropriate. (2 marks)

b. i) Suggest reasons why magnesium gives the nitride, Mg_3N_2, in addition to its oxide   when burned in air. (2 marks)

ii) A 1.00g sample of the powder obtained from burning magnesium in air was boiled with water. The ammonia that was evolved neutralised 12.0 cm^3 of 0.5mol dm^{-3} hydrochloric acid. Write a balanced equation for the reaction that produced magnesium nitride and its reaction with water. (2 marks)

iii) Calculate the percentage of the magnesium nitride in the 1.00g sample. (2 marks)

15.2

a. Magnesium sulphate is readily soluble in water whilst barium sulphate is insoluble.

   i) Suggest an explanation for this observation. (3 marks)

_____

_____

_____

   ii) Describe the trend in thermal stability for the Group 2 nitrates. Explain your answer and include relevant balanced equations where necessary. (5 marks)

_____

_____

_____

_____

_____

b. i) Describe and explain the trend in the melting points of the oxides of the Group 2 elements. (3 marks)

_____

_____

_____

_____

   ii) State a use of one of these oxides based on its melting point. (1 mark)

_____

Chapter 16

THE GROUP IV ELEMENTS AND THEIR COMPOUNDS

Introduction

The elements of Group IV are:

Carbon	C	$[He]2s^22p^2$
Silicon	Si	$[Ne]3s^23p^2$
Germanium	Ge	$[Ar]3d^{10}4s^24p^2$
Tin	Sn	$[Kr]4d^{10}5s^25p^2$
Lead	Pb	$[Xe]4f^{14}5d^{10}6s^26p^2$

Group IV is possibly the most interesting Group in the Periodic Table in the way that trends and patterns between the elements occur. In the other Groups, the elements all have similar properties, and certain trends in these are apparent but in Group IV the elements are very different from one another. The most striking difference is the change from non-metallic carbon at the top of the Group, to metallic tin and lead at the bottom. In between silicon and germanium are known as metalloids because their properties fall between those of non-metals and those of metals.

This trend shows itself in three main ways:
- Carbon has a giant atomic structure in diamond and graphite, whereas lead has a typical giant metallic close-packed structure.
- The oxides of carbon, CO_2 and silicon, SiO_2 are acidic, which is a typical feature of non-metals, whereas the oxides of tin, SnO_2 and lead, PbO_2 are amphoteric, that is, they behave as both acids and bases.
- The compounds of carbon and silicon are covalent, typical of non-metals, whereas tin and lead form ionic compounds containing Sn^{2+} and Pb^{2+} ions, typical of metals.

The physical properties of group 4 elements are summarized in Table 16.1 below.

Element	C	Si	Ge	Sn	Pb
Atomic radius/nm	0.077	0.118	0.122	0.140	0.154
First ionisation energy/kJmol^{-1}	1086	789	762	709	716
Structure	Giant atomic ←		→	giant metallic	→
Electrical conductivity	None (diamond) (graphite)	← semiconductor →		good	good
Melting point/°C	3652 (graphite)	1410	937	232	328
Boling point/°C	4827	2355	2830	2270	1740

Table 16.1 *Physical properties of Group IV elements*

Melting points and electrical conductivities

The melting points of the Group IV elements decrease on descending the Group. The electrical conductivity of these elements increases on descending the group. Both these are explained by the type of structure and bonding of each element.

At the top of the Group, carbon has a giant atomic structure in diamond. There is a very dense network of strong covalent bonds, which results in the melting point being very high. Diamond does not conduct electricity as there are no mobile electrons since all are firmly held in covalent bonds.

At the bottom of the Group, tin and lead have a giant metallic structure. The outer p^2 electrons are held less tightly, as the distance from the nucleus and shielding are both increased, so they form a delocalized sea of electrons. The melting point is lower because heat energy can be transferred through the metal lattice fairly easily, and the electrical conductivity is good, as expected for a typical metallic structure.

Silicon and germanium are semiconductors – they conduct electricity only under certain conditions.

Uses of some Group 4 elements

Carbon, in the form of diamond, is used in jewellery and is also useful in industry because it is very hard. Diamond forms the tips of bits used for drilling rocks. It is also used for cutting glass and other diamonds.

Carbon in the form of graphite is used as a lubricant, especially in heavy-duty machines and as the 'lead' in pencils. It is also used as an inert electrode in electrolysis.

Silicon is used as a semiconductor in the electronics industry, in the form of silicon chips. Silicon is also a component of glass and lubricants. Silicones are compounds that are used in plastic and cosmetic surgery. Silicates are used as the basis of ceramics including glassware, roofing and floor tiles, parts of engines and temperature-resistant furnaces.

Tin is used as a surface layer to protect iron and steel object from rusting, such as tin cans.

Lead is a dense, malleable, soft metal, which makes it useful for electrodes in car batteries. It is also used as a screen against radiation in hospital X-ray departments because of its high density.

Oxidation states and bonding

The elements of Group IV have four outer-shell electrons, two electrons in the s subshell and two electrons in the p subshell, s^2p^2. They all form compounds in which they have oxidation number **+4**, such as CCl_4, tetrachloromethane and SiO_2, silica. In these compounds all four of the outer-shell electrons take part in **covalent bonding**.

Tin and lead also occur in compounds with oxidation number **+2**, such as $SnCl_2$, tin (II) chloride and $PbSO_4$, lead (II) sulphate. This shows that the lower +2 oxidation state becomes more stable as the group is descended. The reason for this is that there is an increasing tendency for the two, s electrons, not to take part in the bonding, as the atomic size increases. This is known as the **inert-pair effect**. The two, p electrons are attracted less strongly to the nucleus, as the atomic size and shielding increase. They can therefore be easily lost to form an ion with a charge of 2+ and give the bonding characteristics of a metal, that is, **ionic bonding**. Metals are generally good reducing agents, and the reducing powers of tin and lead can be seen from their standard electrode potentials.

$E^{\varnothing}$ $Sn^{2+}/Sn =$ -0.14V and $E^{\varnothing}$ $Pb^{2+}/Pb =$ -0.13 V

Both values are negative, which means that both metals can displace hydrogen from acids.

Expansion of the octet

Normally the elements of Group IV have a valence of 4 in their compounds but they can all, except carbon, make more than four bonds. This is because they are able to use a set of d orbitals in their bonding. The d orbitals they use are the empty ones of the outer shell. The 3d set for silicon, the 4d set for germanium and so on. By using these orbitals, the element can hold more than eight electrons in its bonding shell. This is called the **expansion of the octet**. It cannot happen in carbon because there are no, d orbitals available in the second shell of carbon, which is its outermost shell.

In the complex ion SiF_6^{2-}, there is an expansion of the octet. The silicon atom provides four bonding electrons for four fluorine atoms, in the 3s and 3p orbitals. The two fluoride ions occupy the 3d orbitals. They are bonded by coordinate bonding and provide two extra electrons, which give the complex an overall charge of 2^-.

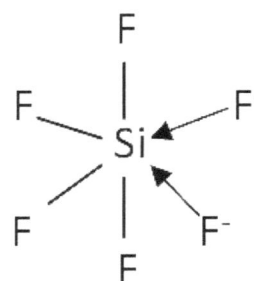

Fig. 16a *the expansion of the octet in SiF_6^{2}*

The properties of the tetrachlorides of Group IV elements

These compounds have the general formula ECL_4 (E being the Group IV element). They have a simple molecular structure, with strong covalent bonds between the atoms. The molecules are tetrahedral in shape (Figure 16b). All the tetrachlorides are **volatile liquids** at room temperature. Carbon tetrachloride, also called tetrachloromethane, has a pungent smell and is used as a dry-cleaning solvent for non-polar stains.

- They have low melting and boiling points because the molecules are non-polar, so there are weak temporary dipole – temporary dipole forces (Van der Waal forces) holding the molecules together in the solid and liquid states.
- The **thermal stability** of the tetrachlorides decreases on descending the Group, as the E - Cl bond gets longer and weaker, due to the increase in the atomic radius of the Group 4 element. On descending the Group compounds with oxidation number +2 become more stable. CCl_4, $SiCl_4$, $GeCl_4$ are stable at high temperatures, while $SnCl_4$ decomposes on heating:

    $SnCl_{4(l)} \rightarrow SnCl_{2(s)} + Cl_{2(g)}$

    and $PbCl_4$ decomposes near room temperature:

    $PbCl_{4(l)} \rightarrow PbCl_{2(s)} + Cl_{2(g)}$

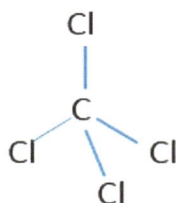

Fig. 16b The tetrahedral shape of carbon tetrachloride.

Reaction of the tetrachlorides with water

All the tetrachlorides except CCl_4 are readily **hydrolysed** in water. For example, silicon tetrachloride is immediately converted to silica, SiO_2.

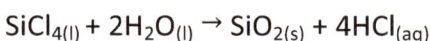

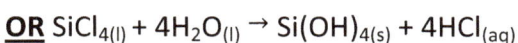

$$SiCl_{4(l)} + 2H_2O_{(l)} \rightarrow SiO_{2(s)} + 4HCl_{(aq)}$$

OR $SiCl_{4(l)} + 4H_2O_{(l)} \rightarrow Si(OH)_{4(s)} + 4HCl_{(aq)}$

In these reactions hydrochloric acid, $HCl_{(aq)}$, is always formed. However, carbon tetrachloride is completely unaffected by water. The explanation for this difference between CCl_4 and the rest of the tetrachlorides, lies in their electronic structure. Carbon does not have a d-subshell while the other elements of Group 4, do. When a water molecule attacks silicon tetrachloride (Fig. 16c), a lone pair on the oxygen atom is attracted to the silicon atom. This is because the silicon atom in silicon tetrachloride has a partial positive charge, due to the electronegative chlorine atoms withdrawing electron density. The chlorine atoms therefore have a partial negative charge. For a short while, the silicon atom has five bonds around it, the extra bond being placed in a 3d orbital, hence silicon can expand its octet. This process is known as **hydrolysis.** However, carbon cannot expand into a d-orbital, so a water molecule is unable to bond to the carbon atom in CCl_4, and hence CCl_4 cannot be hydrolysed.

As the silicon tetrachloride is hydrolysed, O – H and Si – Cl bonds weaken and break, releasing $H^+_{(aq)}$ ions and $Cl^-_{(aq)}$ ions in solution.

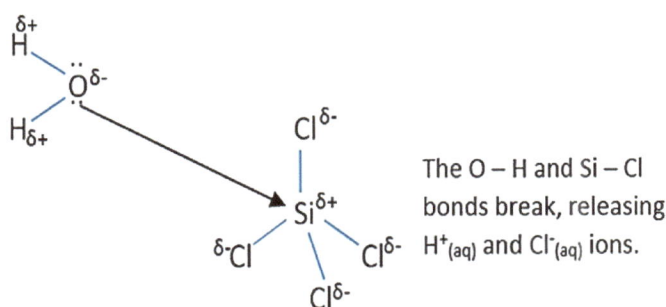

The O – H and Si – Cl bonds break, releasing $H^+_{(aq)}$ and $Cl^-_{(aq)}$ ions.

Fig. 16c *The reaction of water with silicon (iv) chloride*

145

The oxides of the Group IV elements

Oxide of +4 oxidation state	CO_2	SiO_2	GeO_2	SnO_2	PbO_2
Structure	Simple molecular	Giant molecular	Intermediate between giant molecular and giant ionic		
Acid-base nature	acidic	acidic	amphoteric	amphoteric	amphoteric
Reactions	React with alkalis $CO_2 + 2OH^- \rightarrow CO_3^{2-} + H_2O$ carbonate $SiO_2 + 2OH^- \rightarrow SiO_3^{2-} + H_2O$ silicate		React with acids and alkalis Acids must be concentrated: $SnO_2 + 4H^+ \rightarrow Sn^{4+} + 2H_2O$ Alkalis must be molten: $PbO_2 + 2OH^- \rightarrow PbO_3^{2-} + H_2O$ plumbate		
Thermal stability	Stable even at high temperatures				Decomposes to PbO on warming $2PbO_2 \rightarrow 2PbO + O_2$

Table 16.2 *Properties of dioxides of group IV elements – oxidation number of +4*

Oxide of +2 oxidation state	CO	SiO	GeO	SnO	PbO
Structure	Simple molecular	Simple molecular	⟵ giant ionic lattice ⟶		
Acid-base nature	neutral	neutral	amphoteric	amphoteric	amphoteric
Reactions	Do not react with acids or alkalis		React with aqueous acids and alkalis to form salts $SnO_{(s)} + 2H^+_{(aq)} \rightarrow Sn^{2+}_{(aq)} + H_2O_{(l)}$ $PbO_{(s)} + OH^-_{(aq)} + H_2O_{(l)} \rightarrow Pb(OH)_3^-$ (trihydroxyplumbate)		
Thermal stability	Readily oxidised to dioxide				Stable

Table 16.3 *Properties of monoxides of Group IV elements – oxidation number of +2*

The oxides of Group 4 have very different structures, ranging from simple molecular carbon dioxide gas to the giant ionic lead(II)oxide.

Carbon dioxide and silicon dioxide are acidic. Both react with hot concentrated alkalis in aqueous solution, or with solid pellets of potassium hydroxide or sodium hydroxide to form a salt and water.

$SiO_{2(s)} + 2OH^-_{(aq)} \longrightarrow SiO_3^{2-}_{(aq)} + H_2O_{(l)}$ (silicate salt)

Silicon (IV) oxide also reacts with metal carbonates and displaces carbon dioxide.

$SiO_{2(s)} + Na_2CO_{3(s)} \longrightarrow Na_2SiO_{3(s)} + CO_{2(g)}$
Sodium silicate

As the group is descended, the oxides become more basic in character. The oxides of germanium and tin are amphoteric rather than basic, and react with both acids and alkalis.

$SnO_{2(s)} + 2NaOH_{(aq)} \longrightarrow Na_2SnO_{3(aq)} + H_2O_{(l)}$ (stannate IV salt)

$SnO_{2(s)} + 4HCl_{(aq)} \longrightarrow SnCl_{4(aq)} \longrightarrow SnCl_2 + Cl_2 + 2H_2O_{(l)}$ (with concentrated acid)

$PbO_{2(s)} + 4HCl_{(aq)} \longrightarrow PbCl_{2(s)} + Cl_{2(g)} + 2H_2O_{(l)}$ (conc. HCl, **room temp**).

(red-brown)   (white)

Melting points of the dioxides of Group 4 elements

Oxide	Structure	Bonding	Melting point /°C
CO_2	simple molecular	Van der Waal forces	-56
SiO_2	giant molecular	covalent bond	1610
GeO_2	giant molecular	covalent bond	1116
SnO_2	giant molecular	covalent bond	1630
PbO_2	giant molecular	covalent bond	Decomposes at 290

Table 16.4 The melting points of the dioxides of Group IV elements

These melting points reflect the great difference in the bonding of solid carbon dioxide, made up of simple molecules with weak temporary dipole-temporary dipole forces between them, and the other dioxides which are all giant molecular structures with strong covalent bonds throughout.

Tin (II) oxide, SnO, lead (II) oxide, PbO are predominantly ionic but are amphoteric in nature.

They react with acids to form simple salts, and with alkalis to form complex salts.

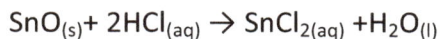

$SnO_{(s)} + 2HCl_{(aq)} \rightarrow SnCl_{2(aq)} + H_2O_{(l)}$

$PbO_{(s)} + NaOH_{(aq)} + H_2O_{(l)} \rightarrow NaPb(OH)_{3(aq)}$ sodium trihydroxyplumbate

Exercise 16.1
a. Describe and explain in terms of bonding and structure, the trend in electrical conductivities and melting points of the Group 4 elements. (4 marks)

_____

_____

_____

_____

b. The tetrachlorides of Group 4, except carbon tetrachloride, produce an acidic solution in water. Explain this difference in behaviour. (3 marks)

_____

_____

_____

_____

c. Triplumbic tetroxide, Pb_3O_4, when treated with dilute nitric acid, produces a brown solid 'A' and a colourless solution 'B'. When 'A' was treated with concentrated hydrochloric acid, it produced a yellowish-green gas 'C' and a colourless solution, which produces a white precipitate, 'D' on cooling.

i) Identify A, B, C and D (4 marks)

_____

_____

_____

_____

ii) Write equations for all the reactions that occur. (2 marks)

_____

_____

_____

_____

_____

Chapter 17

THE GROUP VII ELEMENTS AND THEIR COMPOUNDS

The elements of Group VII are called the halogens:

Fluorine   F   [He]$2s^2 2p^5$
Chlorine   Cl   [Ne]$3s^2 p^5$
Bromine   Br   [Ar]$3d^{10} 4s^2 4p^5$
Iodine   I   [Kr]$4d^{10} 5s^2 5p^5$
Astatine   At   [Xe]$4f^{14} 5d^{10} 6s^2 6p^5$

All the isotopes of astatine are radioactive and so this element will not be considered here. Fluorine shall also not be included in all discussions on Group VII, because its small size and high electronegativity give it some anomalous properties.

The halogens are a family of non-metallic elements with similar chemical properties. Their reactivity decreases down the Group. Their chemical reactions are characterised by the outermost 7 electrons. The most common oxidation state for the halogens is -1, although chlorine exhibits a range of oxidation numbers from -1 to +7. The halogen elements are diatomic molecules with a simple molecular structure.

Fluorine, chlorine and bromine are poisonous. Melting and boiling points increase down the group with increasing relative molecular mass. Fluorine and chlorine are gases at room temperature, bromine is a liquid and iodine is a solid. Volatility decreases down the group as a result of increasingly strong Van der Waals forces as the atomic number and relative molecular mass increase down the group.

The colour of the elements deepens with increasing atomic number and density: fluorine is a pale yellow gas; chlorine is a greenish yellow gas; bromine is a dark red liquid giving off a dense red vapour; iodine is a shiny, black, crystalline solid which sublimes to a purple vapour.

The standard electrode potentials of the halogens, $E^{\varnothing} X_2/2X^-$, range from +2.87V for fluorine, to +0.54V for iodine. This shows that they are all **oxidizing agents**, with fluorine being the strongest. The oxidizing ability is reflected by their reactivity and by their electronegativities. Fluorine is the most reactive halogen, and it is the most electronegative element.

The smaller the atomic radius, the more strongly incoming electrons can be attracted to the atom. Hence as the group is descended and the atomic radius increases, the electronegativity and reactivity of the halogens decrease.

Summary of General Properties of Group VII elements
The general properties of the Group VII elements chlorine, bromine and iodine are as follows:
- They behave chemically in a similar way.
- They are non-metals.
- They all exist as diatomic molecules, X_2, at room temperature.
- Their melting and boiling points increase down the group, with increasing atomic number as the Van der Waal forces increase in strength.

- The colour of the elements deepens with increasing atomic number and as the density increases.
- They are very reactive and readily form salts.
- In compounds, a halogen atom increases its share of electrons from seven to eight by ionic or covalent bonding.
- The reactivity of the elements decreases on descending the Group as the atomic radius increases.
- They exhibit a range of oxidation numbers.
- The electronegativity of the elements decreases on descending the Group.
- Their oxidizing ability decreases on descending the Group.

Physical properties of Group VII elements

Element	F	Cl	Br	I
Atomic radius/nm	0.071	0.099	0.114	0.133
Ionic radius/nm	0.133	0.180	0.195	0.215
Electronegativity	4.0	3.0	2.8	2.5
Electron affinity/ kJ mol^{-1}	-328	-349	-325	-295
Melting point/°C	-220	-101	-7	114
Boiling point/°C	-188	-35	59	184

Table 17.1 *Physical properties of Group VII elements.*

The low melting and boiling points of the solid and liquid halogens is due to the weak Van der Waal forces between the molecules, the strength of which increases, as the atomic number / number of electrons increases down the Group.

Uses of some Group VII elements and their compounds

The oxidizing ability of the halogens means that they are useful in many ways. **Chlorine** and its aqueous solution, known as chlorine water, are often used as oxidizing agents. Chlorine water contains hydrochloric acid, HCl, and chloric (I) acid, HClO. Chlorine is used commercially and domestically as a bleach, oxidising coloured stains to colourless compounds. Very white paper pulp can be produced by bleaching with chlorine. The process results in the formation of dioxins, which are poisonous and they can accumulate in living organisms. Dioxins do not break down easily. Today, ozone is often used to bleach paper that does not have to be pure white, like tissues and toilet paper.

The strong oxidizing ability of chlorine is also used by the water industry to treat drinking water. Chlorine is added to water in reservoirs to kill bacteria and other pathogens. Small amounts of chlorine remain in the water piped to consumers to prevent bacterial contamination. Chlorine is also used to keep water in swimming pools free from contamination.

One of the classes of organic chemicals made, using chlorine is that of CFCs, chlorofluorocarbons. CFCs are used as aerosol propellants, refrigerants and as foaming agents in polymers. They are currently being withdrawn from many applications because they are pollutants and are believed to contribute to the destruction of the ozone layer. However, they are useful in at least two ways. They are used in fire extinguishers because they are inert and non-flammable, and they are vital constituents of artificial blood.

Solvents containing chlorine, such as tetrachloromethane, CCl_4, are widely used to dissolve fats and oils, and are used in the dry cleaning industry.

In the First World War, chlorine and mustard gas ($ClCH_2CH_2SCH_2CH_2Cl$) were used with devastating effect as poison gases. Chlorine is produced by the electrolysis of brine, conc. NaCl.

Fluorine is used, like chlorine, in CFCs. It is also used to make **PTFE, polytetrafluoroethene**, which is used as a lubricant, and as a coating for non-stick cooking pans, electrical insulation and in water-proof clothing.

Fluoride ions help to prevent tooth decay. Some children are given fluoride tablets; many tooth pastes contain tin fluoride (SnF_2); and some water supplies are fluoridated with sodium fluoride. **Hydrofluoric acid, HF** is used to etch glass.

Bromine is used in 1,2-dibromoethane, $BrCH_2CH_2Br$, which is a petrol additive.

Bromochloromethane, CH_2ClBr is used in fire extinguishers.

Silver bromide is used in photographic film.

Iodine is an essential part of our diet, and an imbalance can cause thyroid problems. A solution of iodine in alcohol is sometimes used as an antiseptic.

The reactivity of the halogens

Displacement reactions

The decreasing reactivity of the halogens chlorine, bromine and iodine is shown by their decreasing standard electrode potentials (Table 17.2 below).

Element	$E^{\varnothing} X_2/2X^-/V$
Chlorine	+1.36
Bromine	+1.09
Iodine	+0.54

Table 17.2 *Standard electrode potentials of Group VII elements in contact with solutions of their ions*

The $E^{\varnothing}$ values are for the process:

$$X_{2(aq)} + 2e^- \rightarrow 2X^-_{(aq)}$$

In most of their oxidizing reactions, the halogens react as X_2 molecules and form hydrated halide ions, $X^-_{(aq)}$. As the oxidizing ability decreases from chlorine to iodine, any halogen can displace another with a less positive standard electrode potential. This means that, if each halogen is reacted with a halide ion in aqueous solution, a series of displacement reactions occurs.

Therefore chlorine will displace bromine and iodine as follows:

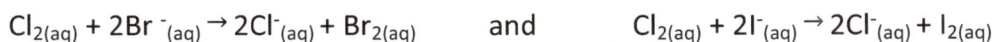

$$Cl_{2(aq)} + 2Br^-_{(aq)} \rightarrow 2Cl^-_{(aq)} + Br_{2(aq)} \quad \text{and} \quad Cl_{2(aq)} + 2I^-_{(aq)} \rightarrow 2Cl^-_{(aq)} + I_{2(aq)}$$

Bromine will displace iodine: $Br_{2(aq)} + 2I^-_{(aq)} \rightarrow 2Br^-_{(aq)} + I_{2(aq)}$

Iodine does not displace either chlorine or bromine but would displace astatine

One of the problems with doing these displacement reactions is being able to see if a reaction has taken place. The halide ion solutions are all colourless and very dilute solutions of the halogens can also appear colourless. To avoid this problem, an organic solvent such as hexane or cyclohexane is added to the mixture, which forms a separate layer. The halogens, being non-polar, are more soluble in organic solvents than in aqueous solution, so they are taken up by the hexane and the colour is much more apparent. For instance, bromine is a strong reddish-brown colour in hexane, and iodine is purple. So, if aqueous bromine is mixed with hexane, the bromine dissolves in the hexane, which turns reddish-brown/orange. Then if aqueous potassium iodide is added, the hexane turns purple, which shows that bromine has displaced iodine from solution.

Halide			
Halogen	**Chloride, Cl⁻**	**Bromide, Br⁻**	**Iodide, I⁻**
Chlorine, Cl_2		reddish-brown bromine released in organic solvent	purple iodine released in organic solvent. Example hexane
Bromine, Br_2	no reaction		purple iodine released
Iodine, I_2	no reaction	no reaction	

Table 17.3 *displacement reactions of halogens (colour refer to the colours of the halogens in cyclohexane)*

The Reactions of the halogens with thiosulphate ions

All the halogens react with the thiosulphate ion, $S_2O_3^{2-}$. The thiosulphate ion is a reducing agent. Chlorine and bromine produce sulphate ions.

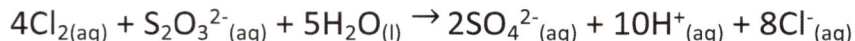

$$4Cl_{2(aq)} + S_2O_3^{2-}{}_{(aq)} + 5H_2O_{(l)} \rightarrow 2SO_4^{2-}{}_{(aq)} + 10H^+_{(aq)} + 8Cl^-_{(aq)}$$

Because chlorine is a strong oxidising agent, it oxidizes the thiosulphate ion completely, to the sulphate ion. There is an increase from +2 to +6 in the oxidation state of sulphur.

However, because Iodine is a weaker oxidising agent, it only partially oxidizes the thiosulphate to the tetrathionate ions, $S_4O_6^{2-}{}_{(aq)}$, an increase from +2 to +2.5 in the oxidation state of sulphur.

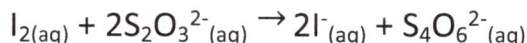

$$I_{2(aq)} + 2S_2O_3^{2-}{}_{(aq)} \rightarrow 2I^-_{(aq)} + S_4O_6^{2-}{}_{(aq)}$$

Reaction of the halogens with hydrogen: The hydrogen halides

Hydrogen chloride, hydrogen bromide and hydrogen iodide can be prepared in this way:

$$H_{2(g)} + X_{2(g)} \rightarrow 2HX_{(g)}$$

The production of the hydrogen halides is summarised in Table 17.4 below.

The hydrogen halides are acids. The covalent bond strengths and bond energies decrease in the order H-F > H-Cl > H-Br > H-I as the halogen atom increases in size. (Table 17.5)

HF, hydrofluoric acid therefore donates a proton least readily and is a weak acid. HI, hydroiodic acid is the strongest of the hydrohalic acids.

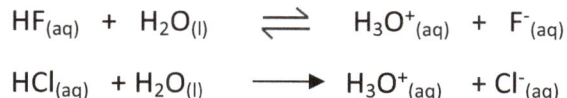

$$HF_{(aq)} + H_2O_{(l)} \rightleftharpoons H_3O^+_{(aq)} + F^-_{(aq)}$$

$$HCl_{(aq)} + H_2O_{(l)} \longrightarrow H_3O^+_{(aq)} + Cl^-_{(aq)}$$

How does each halogen react with hydrogen?

Decreasing reactivity down the Group	chlorine	Explodes in direct sunlight. Slow reaction in the dark.
	bromine	Reacts at 300°C and using a platinum catalyst.
	iodine	Slow reaction at 300°C and using a platinum catalyst. Easily reversible, so a partial product obtained.

Table 17.4 *Reactions of halogens with hydrogen*

$$H_{2(g)} + Cl_{2(g)} \longrightarrow 2HCl_{(g)} \quad : \quad H_{2(g)} + I_{2(g)} \rightleftharpoons 2HI_{(g)}$$

Hydrogen Halide	Bond dissociation enthalpies $\Delta H^{\varnothing}_{dissociation(HX)}$ kJmol^{-1}
HCl	432
HBr	366
HI	298

Table 17.5 *Bond dissociation enthalpies for hydrogen halides*

The **thermal stability** of the hydrogen halides decreases in the order: HCl > HBr > HI

On descending the Group, the atomic radius of the halogen increases, resulting in weaker covalent bond between the hydrogen atom and the halogen. As the covalent bond in the hydrogen halides, gets weaker down the Group, less energy is required for decomposition.

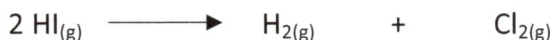

$$2 HI_{(g)} \longrightarrow H_{2(g)} + Cl_{2(g)}$$

However, the **acid strength** of these compounds increases in the order: HCl < HBr < HI

This is because hydrogen iodide, with the weakest covalent bond, will donate its proton most readily in aqueous solution. This makes hydroiodic acid, the strongest of the hydrohalic acids. Hydrofluoric acid is the weakest.

$$HCl_{(aq)} + H_2O_{(l)} \longrightarrow H_3O^+_{(aq)} + Cl^-_{(aq)}$$

Confirmatory tests for the halide ions

The reagents used to test for halides are: (i) a solution of silver nitrate, (ii) dilute ammonia solution and, (iii) concentrated ammonia solution.

This test is based on the colour of the silver halide precipitates, and their different solubilities in ammonia solution.

Method and Observations

- Acidify the unknown halide solution with dilute nitric acid
- Add silver nitrate solution to precipitate the silver halide:
  - Silver chloride is a white precipitate
  - Silver bromide is a cream precipitate
  - Silver iodide is a yellow precipitate
- Identification by colour is not completely reliable, so add ammonia solution:
  - The white silver chloride precipitate is completely soluble in dilute ammonia solution but the silver bromide is only slightly soluble and silver iodide precipitate is insoluble.

  - The cream silver bromide precipitate as well as the silver chloride precipitate, are soluble in concentrated ammonia solution but the yellow silver iodide precipitate is insoluble.

Concentrated sulphuric acid and the halide ions

Concentrated sulphuric acid is an **oxidizing agent**, and halide ions are **reducing agents**. Since the halogens X_2 are oxidizing agents then it follows that the halides X^- must be reducing agents.

Cl_2 is a strong oxidising agent therefore Cl^- is a weak reducing agent. However, I^- is a strong reducing agent because I_2 is a weak oxidising agent.

When concentrated sulphuric acid and halide ions react together, a different reaction occurs for each ion depending on the strength of the reducing ability of the halide ion.

When potassium **chloride** is reacted with concentrated sulphuric acid, hydrogen chloride is produced. **No redox** reaction occurs because the chloride ion is a very weak reducing agent.

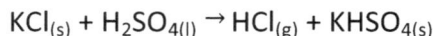

$$KCl_{(s)} + H_2SO_{4(l)} \rightarrow HCl_{(g)} + KHSO_{4(s)}$$

When the **bromide ion, Br⁻** is reacted with the concentrated sulphuric acid, a **redox** reaction takes place. The Br⁻ ion, which is a stronger reducing agent than Cl⁻ ion, reduces the conc. H_2SO_4 to **sulphur dioxide** as the oxidation state of sulphur decreases from +6 to +4. The Br⁻ ion is itself oxidized to bromine:

$$2HBr_{(g)} + H_2SO_{4(l)} \rightarrow Br_{2(l)} + SO_{2(g)} + 2H_2O_{(l)}$$

The **iodide ion, I⁻** is the strongest reducing halide ion and it reduces concentrated sulphuric acid further to **hydrogen sulphide** in which the oxidation state of sulphur decreases from +6 to -2.

$$8HI_{(g)} + H_2SO_{4(l)} \rightarrow 4I_{2(s)} + H_2S_{(g)} + 4H_2O_{(l)}$$

Hydrogen sulphide, H_2S can be recognized by its 'rotten egg' smell and by its black precipitate of lead sulphide with lead ethanoate.

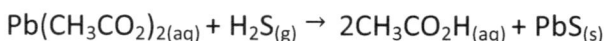

$$Pb(CH_3CO_2)_{2(aq)} + H_2S_{(g)} \rightarrow 2CH_3CO_2H_{(aq)} + PbS_{(s)}$$

The reaction of chlorine with sodium hydroxide

The way in which chlorine reacts with aqueous sodium hydroxide depends on the temperature.

At 15°C, using cold dilute aqueous sodium hydroxide, a mixture of halide (Cl⁻) and halate (I)

(ClO⁻) ions is formed:

$Cl_{2(g)} + 2NaOH_{(aq)} \rightarrow NaCl_{(aq)} + NaClO_{(aq)} + H_2O_{(l)}$

This is a redox reaction called a **disproportionation reaction** in which the same species is oxidized and reduced at the same time.

This happens to the chlorine in the ionic equation below. The oxidation number of chlorine decreases from 0 to -1 in one product, Cl^- and increases from 0 to +1 in the other product, ClO^-.

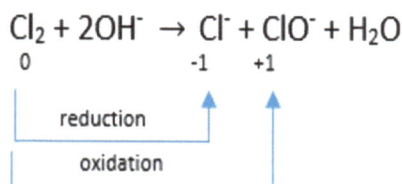

$$Cl_2 + 2OH^- \rightarrow Cl^- + ClO^- + H_2O$$

At 70°C using hot aqueous sodium hydroxide, the ClO^- ion disproportionates further to produce the chlorate (V) ion:

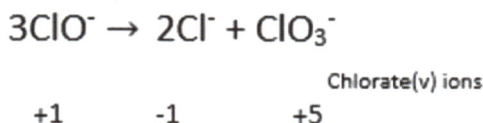

$$3ClO^- \rightarrow 2Cl^- + ClO_3^-$$

The entire equation is:

$3Cl_{2(g)} + 6NaOH_{(aq)} \rightarrow 5NaCl_{(aq)} + NaClO_{3(aq)} + 3H_2O_{(l)}$

Exercise 17.1

a. i) Describe and explain the trend in volatility of the halogens down the Group. (2 marks)

_____

_____

   ii) Describe the reactions of the four halogens fluorine to iodine with hydrogen. (4 marks)

_____

_____

_____

iii) Write balanced equations for the reaction with chlorine and iodine. (2 marks)

_____

_____

b. Describe and explain the trend in the relative stabilities of the hydrides of Group 7. (3 marks)

_____

_____

_____

c. i) Which halogen is the strongest oxidising agent? Explain your answer in terms of reactivity with the thiosulphate ion. (3 marks)

_____

_____

_____

ii) State the trend in reducing strength in the halide ions. Illustrate your answer using a suitable oxidising agent. Include relevant equations. (5 marks)

_____

_____

_____

17.2

a. Bromine is obtained from sodium bromide by,

1. passing chlorine through it at pH 3.5,
2. blowing out the bromine with air and absorbing it in aqueous sodium carbonate, and
3. acidifying the solution and distilling out the bromine
   i) Write an equation for the reaction in Step I. (1 mark)

_____

ii) Step II produces a solution of sodium bromate (V) and sodium bromide in a 1:5 mole ratio. Construct an equation for the formation of bromine in step III. (2 marks)

_____

_____

b. Describe the physical states and the colours of chlorine, bromine and iodine at room temperature and explain their trend in volatility. (5 marks)

_____

_____

_____

_____

c. By using the reaction between the chlorate (I) ion and aqueous sodium hydroxide at 70°C, explain the meaning of the term, disproportionation. (2 marks)

_____

_____

_____

Chapter 18

TRANSITION METAL CHEMISTRY

A transition metal is defined as an element that forms at least one ion or compound with a partially filled, d sub-shell. As such they are referred to as the **d block** elements. The elements of Groups 1 and 2 are referred to as the 's' block elements and Groups 3 to 7 as the 'p' block.

Some of the ways in which the 'd' block elements differ from the 's' and 'p' block are as follows:

- Based on the Aufbau Principle, electrons in the ground state occupy the 4s orbital before occupying the 3d orbitals.
- Each additional electron entering the penultimate d subshell provides shielding of the outer 4s electrons, from the nucleus. This nullifies to a great extent, the attractive force of each additional proton from scandium to zinc. The effective nuclear charge and the atomic radius therefore changes only slightly across the series scandium to zinc.
- The energy of the 3d and 4s orbitals is close enough for the 3d orbitals to be degenerate with the 4s orbital.
- Transition metals can therefore lose electrons from both the 4s and 3d orbitals, resulting in variable oxidation states.

Electronic structure of the transition metals

Element	Configuration	3d					4s	Note
^{20}Ca	$1s^2 2s^2 2p^6 3s^2 3p^6$ (Ar)$4s^2$						↑↓	Not a transition metal
^{21}Sc	Ar	↑					↑↓	
^{22}Ti	Ar	↑	↑				↑↓	
^{23}V	Ar	↑	↑	↑			↑↓	
^{24}Cr	Ar	↑	↑	↑	↑	↑	↑	Half-filled 'd' sub-shell is more stable
^{25}Mn	Ar	↑	↑	↑	↑	↑	↑↓	
^{26}Fe	Ar	↑↓	↑	↑	↑	↑	↑↓	
^{27}Co	Ar	↑↓	↑↓	↑	↑	↑	↑↓	
^{28}Ni	Ar	↑↓	↑↓	↑↓	↑	↑	↑↓	
^{29}Cu	Ar	↑↓	↑↓	↑↓	↑↓	↑↓	↑	Full 'd' sub-shell stable
^{30}Zn	Ar	↑↓	↑↓	↑↓	↑↓	↑↓	↑↓	All orbitals are full, not typically transition metal

Fig. 18a *electronic structure of transition metals*

Properties of the transition metals

The transition metals are similar in their chemical and physical properties. In contrast, the elements across the periods, for example, the period from Na to Ar, show a gradation in properties from left to right as the atomic number increases. The relatively small difference in the effective nuclear charge of the transition metals, as the atomic number increases, accounts for the general similarity in their properties. Hence, the atomic radius decreases only very slightly across the series of transition metals and the ionisation energy increases very slightly also.

Physical Properties

Transition metals are:

- hard
- have high density
- are good conductors of heat and electricity
- have very high melting and boiling points

These physical properties are due to the strong metallic bond in the transition elements. The availability of the 3d electrons as well as the 4s electrons in the sea of delocalised electrons, results in the strong metallic bond with a close packing lattice structure.

Scandium and **Zinc** do not have typical transition metal properties. This is because their compounds do not have an incomplete sub-shell of d electrons. When scandium loses 3 electrons, its 3d sub-shell is empty ($3d^0$). When zinc loses 2 electrons, its 3d sub-shell is full ($3d^{10}$). Copper is transitional only in copper (II) ($3d^9$) compounds.

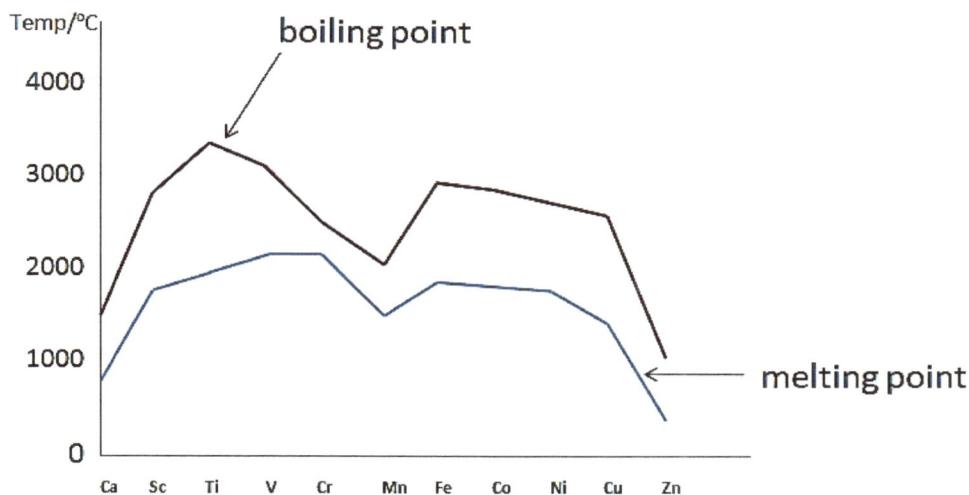

Fig. 18b *boiling points and melting points of transition metals*

The relatively low melting point of manganese is attributed to the stability of the half-filled d sub-shell (d^5) and therefore a lower availability of such electrons for delocalization, hence a relatively weaker metallic bond.

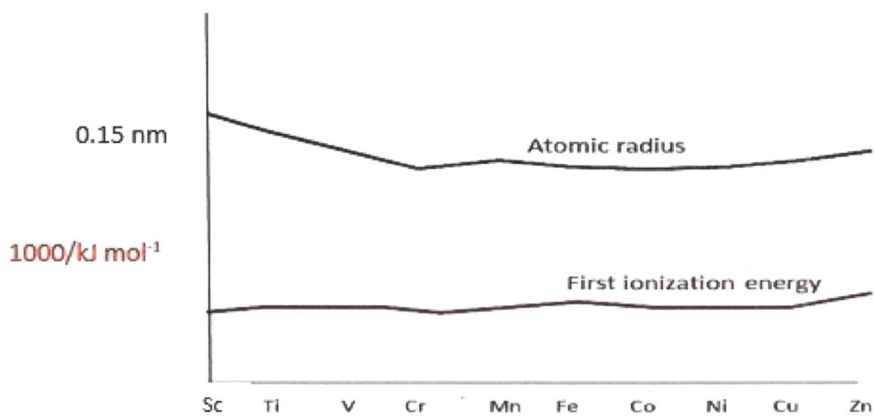

Fig. 18c *Trends in atomic radius and first ionization energy of the transition metals*

First and second ionization energies increase only very slightly and gradually as electrons are added to the d sub-shell. This is due to the minimal increase in the effective nuclear charge from element to element.

Chemical Properties

The transition metals are less reactive than the Group 1 and Group 2 metals and they display much similarity in their chemical reactions.

They form complex ions, have magnetic properties, exhibit variable oxidation states, act as catalysts and form coloured compounds/ions.

Transition metal ions form complexes with electron rich species called **ligands.** They do so by **co-ordinate bonding.** Ligands are electron rich species such as anions and molecules with lone pairs of electrons. As a ligand approaches the transition metal ion during the bonding, there are repulsive forces between the d electrons and the electrons of the approaching ligand. This results in an energy split, **ΔE,** among the five d orbitals and they are **no longer degenerate.**

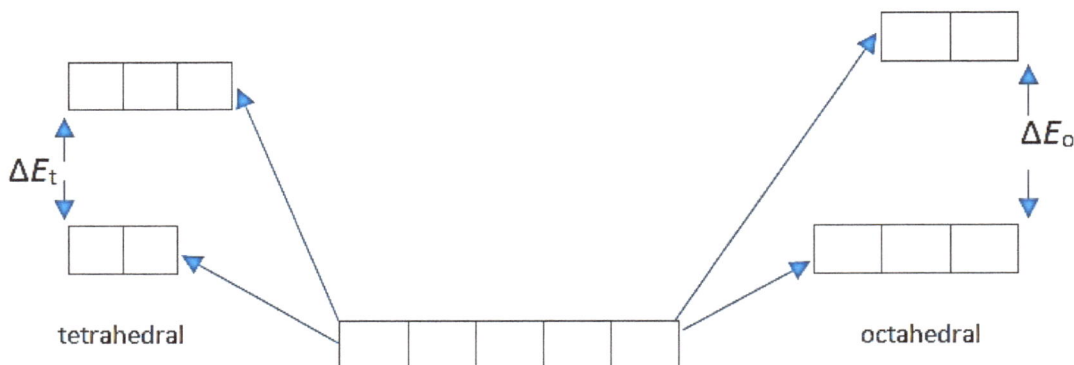

Fig. 18d *The splitting of the five d-orbitals in tetrahedral and octahedral fields*

$Cr_2O_7^{2-}$ and MnO_4^- are examples of complex ions formed from transition metals. During bonding there is an overlap between a lone pair of the ligand and a vacant d orbital of the transition metal ion. Examples of ligands are: NH_3, CN^-, Cl^- and H_2O.

The magnitude of this energy split (ΔE) depends on:

1. **The nature of the ligand**

   Some ligands are more strongly attracted to the transition metal ion and cause a greater energy split. For example, CN^- causes a greater energy split than $NH_3 > H_2O > OH^- > F^- > Cl^-$. CN^- is therefore considered a stronger ligand than NH_3 and Cl^- would be the weakest. A strong ligand will displace a weaker ligand from its complex.
2. **The oxidation state** of the transition metal ion

The higher the oxidation state of the metal ion, the more strongly a ligand will be attracted to it. For example, Cr^{3+} will attract a ligand more strongly, causing greater electron repulsion and a greater energy split than Cr^{2+}.

Properties such as **colour, magnetism and stability of the oxidation states** depend on the magnitude of the energy split.

Colour is due to electrons absorbing visible light and moving from lower energy d orbitals to higher energy d orbitals. The wavelengths of light that are reflected by solids, or transmitted through solutions, give colour to the complex.

Magnetism

When an object is placed in a strong magnetic field, one of the following could occur:

- It could experience no force, in which case it is said to be **non-magnetic.**
- It could be repelled from the magnetic field and is described as being **diamagnetic.**
- It could be attracted to the magnetic field and is described as being **para-magnetic.**

Paramagnetism is associated with unpaired electrons. Most chemical compounds are weakly diamagnetic because all of their electrons are paired, but many transition element complexes are often paramagnetic. This is because they contain **unpaired d electrons**. The strength of the paramagnetism shown by complexes is dependent on the total number of unpaired d electrons.

Number of unpaired electrons	Magnetic moment
1	1.71
2	2.84
3	3.86
4	4.91
5	5.94

Table 18.1 *relationship between the number unpaired electrons and magnetic moment*

An extreme form of paramagnetism is **ferromagnetism**, shown only by the elements iron, cobalt and nickel and by some of their oxides.

Oxidation states and their relative stabilities

The transition elements exhibit different oxidation states. This is because of the close similarity in the energy of the 4s and 3d electrons, hence 3d and 4s electrons can be easily lost. This results in one element forming several different ions, exhibiting different oxidation states. Calcium, in Group 2, has an oxidation number of +2 because once the s, electrons are removed the inner p electrons are in a stable shell and require too much energy for their removal. In contrast, for vanadium, it is energetically easy to remove the first 3 electrons from the s and d orbitals as shown in Table 18.2.

| Element | Successive ionization energies kJmol^{-1} | | | |
	$E_{i(1)}$	$E_{i(2)}$	$E_{i(3)}$	$E_{i(4)}$
Ca	590	1150	4940	6480
V	648	1370	2870	4600

Table18.2 *comparing successive ionization energies of calcium and vanadium*

Element	Oxidation states						
Sc			+3				
Ti	+1	+2	+3	+4			
V	+1	+2	+3	+4	+5		
Cr	+1	+2	+3	+4	+5	+6	
Mn	+1	+2	+3	+4	+5	+6	+7
Fe	+1	+2	+3	+4	+5	+6	
Co	+1	+2	+3	+4	+5		
Ni	+1	+2	+3	+4			
Cu	+1	+2	+3				
Zn		+2					

Table 18.3 *Oxidation states of the first row transition metals*

Table 18.3 above, summarises the known oxidation numbers of the elements scandium to zinc. Those in bold are most relevant in this study but are not necessarily the most common or most stable. In their elemental state, the oxidation number is zero.

Note *that Sc^{3+} has no d electrons ($3d^0$) and does not satisfy the definition of a transition element. Also, Zn^{2+} has a full d sub-shell ($3d^{10}$) and therefore is also not considered a transition metal. This is the reason why the compounds of zinc and scandium are white and not coloured like those of other transition elements. Changes in the oxidation state of the ions are often shown by changes in the colour of the solutions. For example, Cr^{+6} as in $Cr_2O_7^{2-}$ is orange whilst Cr^{3+} is blue-violet (green).*

Oxidation states and colours of different vanadium ions

Ion/Complex	Oxidation State	Colour	$E^{\varnothing}$/V
VO_3^-	+5	yellow	+1.00
VO_2^+	+5	yellow	+1.00
VO^{2+}	+4	blue	+0.34
V^{3+}	+3	green	-0.26
V^{2+}	+2	violet	-1.20

Table 18.4 *Oxidation states and colours of different vanadium ions*

The standard electrode potentials, $E^{\varnothing}$ in table 18.4, show that the vanadium, in the +5 oxidation state, is a relatively strong oxidizing agent.

The oxidizing strength decreases down to the +2 oxidation state.

Reactions with a strong reducing agent will bring about a gradual decrease in the oxidation states of the vanadium ion, and therefore, changes in colour.

Equations for reactions of vanadium ions with a strong reducing agent, zinc are shown below:

1)
$$Zn - 2e^- \rightarrow Zn^{2+} \qquad\qquad E^\varnothing \;\; +0.76V$$

$$2VO_3^- + 8H^+ + 2e^- \rightarrow 2VO^{2+} + 4H_2O \;\; +1.00V$$

Overall equation: $Zn + 2VO_3^- + 8H^+ \rightarrow Zn^{2+} + 2VO^{2+} + 4H_2O$ $E^\varnothing_{cell} = +1.76V$ $\qquad$ feasible reaction

2) $Zn + 2VO^{2+} + 4H^+ \rightarrow 2V^{3+} + 2H_2O + Zn^{2+}$ $\qquad$ $E^\varnothing_{cell} = +1.10V$ $\qquad$ feasible reaction

3) $2V^{3+} + Zn \rightarrow 2V^{2+} + Zn^{2+}$ $\qquad$ $E^\varnothing_{cell} = +0.50V$ $\qquad$ feasible reaction

4) $V^{2+} + Zn \rightarrow Zn^{2+} + V$ $\qquad$ $E^\varnothing_{cell} = -0.44V$ $\qquad$ reaction is not feasible

Name an element that would reduce V^{2+} to V. _____

Transition metal complexes

Transition metal ions form complexes, or coordination compounds, with ligands. They donate pairs of electrons to vacant d orbitals of the ions, during coordinate bonding. They are either anions or neutral molecules with lone pairs.

If the ligand donates one pair of electrons to the ion it is called a monodentate ligand. Polydentate ligands have more than one site that bond to the metal ion.

For example, bidentate ligands donate two pairs of electrons and hexadentate ligands donate six pairs of electrons as shown in Table 18.5.

Type of Ligand	Formula	Name
Monodentate	H_2O	Water
	CO	Carbon monoxide
	S^{2-}	Sulphide ion
	NO_2^-	Nitrite ion
Bidentate	$NH_2CH_2CH_2NH_2$	Ethane- 1,2- diamine
	CO_2^- | CO_2^-	Ethanedioate ion
Hexadentate		Edta (ethylenediaminetetraacetic acid) ion

Table 18.5 *Some common ligands*

Writing the formula of and naming a coordinate complex

In writing the formula of a complex, always start with the central transition metal ion, followed by the ligands, and with the overall charge of the complex at the end.

The overall charge on the complex is simply the individual charges of the transition element ion and the ligands added together.

To name the complex ion, start with the number and name of the ligands in alphabetical order, followed by the transition metal and charge. For example, the names of the following transition metal complexes: $[Ni(CN)_4]^{2-}$ and $[Cr(H_2O)_4Cl_2]^+$ would be tetracyanonickelate(II) and tetraaquadichlorochromium(III) complex respectively.

In these two examples, we know that the ligands are identified as CN^-, Cl^- and H_2O, so we can calculate the **charges on the transition metal ions** as follows: x - 4 = - 2 and x + 0 -2 = +1 to give charges of +2 on Ni^{2+} and +3 for Cr^{3+} respectively.

Note that if the coordination complex is an anion, the suffix **-ate** is added at end of the name of the metal. For example nickelate, cobaltate, cuprate, ferrate(II) and ferrate(III).

Shapes of Complexes

The number of lone pairs of electrons bonded to the transition metal ion is called the **coordination number** of the complex.

This is related to the shape of the complex. There are three main shapes adopted by transition element complexes – **octahedral, tetrahedral** and **square planar**.

An octahedral complex has a coordination number of 6, and the other two shapes both have a coordination number of 4.

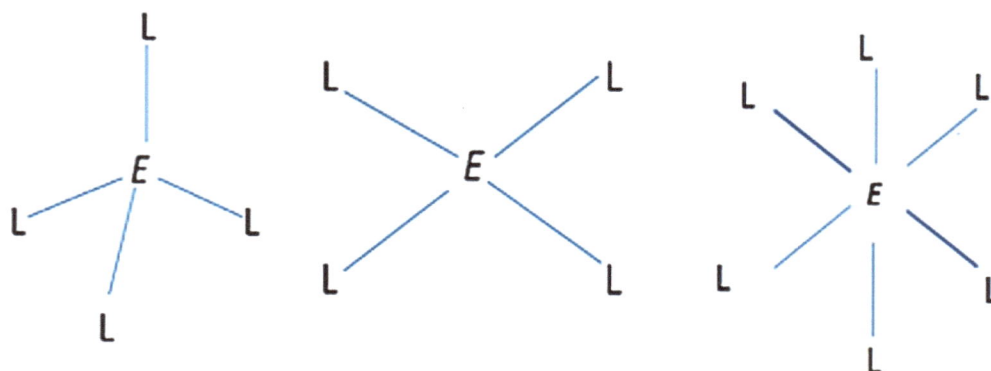

Fig. 18e *shapes of complexes*

One of the best known complexes is formed in a solution of copper(II)sulphate, $CuSO_4$. Solid anhydrous copper(II)sulphate is a white powder. In aqueous solution the copper ion forms a complex with six water molecules. The complex has an octahedral shape and is responsible for the blue colour associated with copper(II)sulphate solution.

Ligand displacement

The water ligands in the copper complex of aqueous copper (II) sulphate can be displaced by a stronger ligand to form a more stable complex. When concentrated hydrochloric acid is added drop by drop, the solution turns yellow, as a new complex is formed. Four water ligands are replaced by four chloride ion ligands to give $[CuCl_4]^{2-}$ as follows:

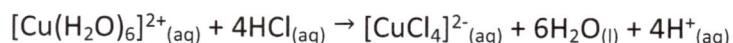

$$[Cu(H_2O)_6]^{2+}{}_{(aq)} + 4HCl_{(aq)} \rightarrow [CuCl_4]^{2-}{}_{(aq)} + 6H_2O_{(l)} + 4H^+{}_{(aq)}$$

Note that only four chloride ligands displace the six water ligands. This is because the chloride ligands are large and the copper(II) ion can 'fit' only four Cl^- ligands compared to six smaller H_2O ligands.

The chloride ion ligands are similarly replaced by ammonia ligands when concentrated ammonia solution is added drop by drop producing a **deep blue** solution of the tetraamminediaquacopper(II) complex:

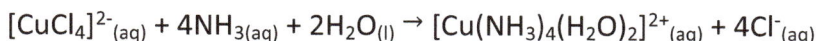

$$[CuCl_4]^{2-}_{(aq)} + 4NH_{3(aq)} + 2H_2O_{(l)} \rightarrow [Cu(NH_3)_4(H_2O)_2]^{2+}_{(aq)} + 4Cl^-_{(aq)}$$

The ammonia ligands can be replaced by edta to give a complex that is turquoise in aqueous solution. This sequence of ligand replacement means that we can list the complexes in order of their stability.

Complex	Log K_{st}	
$[Cu(H_2O)_6]^{2+}$		
$[CuCl_4]^{2-}$	5.6	Increasing stability
$[Cu(NH_3)_4(H_2O)_2]^{2+}$	13.1	
$[Cu(edta)(H_2O)_2]^{2-}$	18.1	

Table 18.6 *The stability of some copper complexes*

Stability constants, K_{st}, like equilibrium constants, express the extent of a reaction. The higher K_{st} for a complex, the more likely it will form. A complexing ligand with a higher K_{st} value will displace one of lower K_{st}. K_{st} is expressed as logK. The larger log K_{st} the more stable the complex.

Carbon monoxide, CO, forms a more stable complex with Fe^{2+} in the haemoglobin of red blood cells, than oxygen. Carbon monoxide therefore 'poisons,' by displacing the oxygen in the oxyhaemoglobin complex of the red blood cells. This results in death from lack of oxygen.

Catalytic activity

Transition elements, both in the elemental form and in their compounds, are effective and important catalysts. A catalyst, which, by definition, alters reaction rate without itself undergoing any permanent chemical change, offers an alternative reaction pathway (one which involves a lower activation energy than that of the uncatalysed process).

Two features of transition elements are important in this regard. One is the **availability of both 3d and 4s electrons**. Reactant molecules can form weak bonds in the vacant d orbitals, on the surface of the catalyst. This occurs in **heterogenous catalysis** and is called adsorption. This adsorption releases energy and weakens the bonds in the reactant molecules, thereby lowering the activation energy. Adsorption also results in increasing the surface concentration of the reactant particles.

The second feature is the **variable oxidation state** of transition metals. This is an important factor in **homogenous catalysis**. For example, in the oxidation of I^- by $S_2O_8^{2-}$ as shown in the equation below:

$$S_2O_8{}^{2-}{}_{(aq)} + 2I^-{}_{(aq)} \rightarrow 2SO_4{}^{2-}{}_{(aq)} + I_{2(aq)}$$

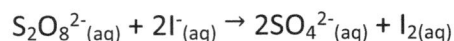

This redox reaction is accelerated by the presence of a number of transition metal catalysts such as Iron (II) and Iron (III) compounds.

The uncatalysed reaction goes to completion but at a slow rate, having a high activation energy. A possible catalysed pathway having lower activation energy in two steps is as follows:

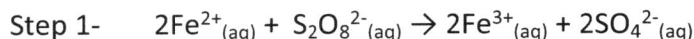

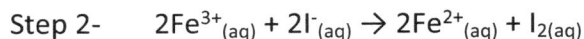

Step 1- $2Fe^{2+}{}_{(aq)} + S_2O_8{}^{2-}{}_{(aq)} \rightarrow 2Fe^{3+}{}_{(aq)} + 2SO_4{}^{2-}{}_{(aq)}$

Step 2- $2Fe^{3+}{}_{(aq)} + 2I^-{}_{(aq)} \rightarrow 2Fe^{2+}{}_{(aq)} + I_{2(aq)}$

The Fe^{2+} ion is regenerated at the end and is a catalyst. The energy of activation is significantly decreased and the rate of the reaction increases.

Colour

Most transition metal compounds are coloured, both in the solid state and in solution. In transition metal complexes, the d orbitals are split into low and high energy orbitals. When **visible light** is passed through a solid, or an aqueous solution of a transition metal complex, radiation is absorbed by unpaired electrons in the low energy d orbitals. As they do so, there is d – d transition as they move into higher energy d orbitals. The reflected or transmitted radiation/light gives the colour of the complex. For example, if there is absorbance in the green, yellow and orange region of the visible spectrum, then red and blue wavelengths are reflected/transmitted, the colour observed will be purple.

Transition metal ion complexes in which all the d orbitals are full or empty, as in Zn^{2+} and Sc^{3+} respectively, cannot absorb visible light. Hence all the white light is reflected/transmitted giving rise to white solids and colourless solutions.

The colour of the complex depends on the following factors:

a) **The number of d electrons present**. The absence of colour in Sc(III), Cu(I) and Zn(II) compounds is due to the fact that they either have no d electrons, d^0 as in (Sc(III)), or a full, d^{10} arrangement as in(Cu(I) and Zn(II)). In either case no d - d transition is possible. Also, depending on the number of d electrons there will be varying oxidation states and colours.

b) **The nature of the ligand**. Different ligands have a different effect on the relative energies of d-orbitals of a particular ion. For example, ammonia ligands cause a larger energy split than water ligands. This factor results in the colour change from blue to blue-violet when ammonia is added to an aqueous solution of a copper (II) salt.

Exercise 18.1

a. i) Define the term 'transition metal'. (1 mark)

_____

_____

b. Write the electronic structure to reflect the 3d and 4s subshells of the following: (5 marks)

ii) Fe

iii) Fe^{2+}

iv) Fe^{3+}

v) Cr

vi) V^{3+}

c. i) Explain why transition element complexes are often coloured. (3 marks)

ii) Explain what determines the colour of a particular transition element complex (2 marks)

18.2

a. Explain the following changes as fully as you can:
   i) The addition of thiocyanate ions to aqueous iron (III) ions produces a deep blood-red colour which disappears when aqueous sodium fluoride is added. (3 marks)

ii) The addition of water to blue anhydrous cobalt (II) chloride produces a pale pink solution. (2 marks)

_____

_____

18.3

a. Many transition metals and their compounds are useful catalysts. By choosing a suitable example in each case, illustrate and explain how iron can act as:

   i) A homogenous catalyst (3 marks)

_____

_____

   ii) A heterogenous catalyst (3 marks)

_____

_____

b. i) What do you understand by the term 'ligand'? (1 mark)

_____

   ii) State, giving a reason, whether or not each of the following molecules or ions can act as a ligand, NH_3, BH_3, Zn^{2+}, I^-, CN^-. (5 marks)

_____

_____

c. The following table lists some stability constants for the following reaction.

$$[M(H_2O)_6]^{m+}{}_{(aq)} + nL_{(aq)} \rightarrow [M(H_2O)_{6-n}L_n]^{(m-n)+}{}_{(aq)} + nH_2O_{(l)}$$

M^{m+}	L^-	n	K_{st}
Fe^{3+}	SCN^-	1	9×10^2
Fe^{3+}	CN^-	6	1×10^{31}
Co^{3+}	CN^-	6	1×10^{64}

i) Rewrite the above equation in which $M^{m+} = Fe^{3+}$ and $L^- = CN^-$. (2 marks)

ii) Write the expression for the equilibrium constant, K_{st} and state its units. (2 marks)

iii) Use the data given in the table to predict what would be the predominant complex formed when:

A solution containing equal concentrations of both SCN^- and CN^- ions are added to a solution containing $Fe^{3+}_{(aq)}$ ions. (1 mark)

A solution containing equal concentrations of $Fe^{3+}_{(aq)}$ and $Co^{3+}_{(aq)}$ ions was added to a solution containing CN^- ions. (1 mark)

Appendix I

Qualitative Analysis

Confirmatory tests for cations

Some reagents that can be used to identify cations are as follows: 1) sodium hydroxide solution, $NaOH_{(aq)}$, 2) ammonia solution, $NH_{3(aq)}$ and 3) sodium carbonate solution, $Na_2CO_{3(aq)}$.

These reagents form precipitates of characteristic colours with different cations.

Sodium hydroxide solution is used as a reagent to test for many cations since most metal hydroxides are insoluble. The cation can be identified by the colour of the precipitate formed and by the solubility of the precipitate in excess alkali.

Ammonia solution also forms precipitates with cations and some will dissolve in excess ammonia solution.

Most carbonates are insoluble, therefore, precipitates are formed with aqueous sodium carbonate. If the carbonate formed is unstable, carbon dioxide gas evolves.

Tests for cations

Cation	Reaction with		
	$NaOH_{(aq)}$	$NH_{3(aq)}$	$Na_2CO_{3(aq)}$
Aluminium, $Al^{3+}_{(aq)}$	White precipitate soluble in excess	White precipitate Insoluble in excess	CO_2 gas evolved
Ammonium, $NH_4^{+}_{(aq)}$	Ammonia produced on heating		$NH_{3(g)}$ evolved
Barium, $Ba^{2+}_{(aq)}$	No precipitate (if reagents are pure)	No precipitate	White precipitate
Calcium, $Ca^{2+}_{(aq)}$	White precipitate with high $[Ca^{2+}_{(aq)}]$	No precipitate	White precipitate
Chromium (III), $Cr^{3+}_{(aq)}$	Grey-green precipitate, soluble in excess giving dark green solution	Grey-green precipitate, insoluble in excess	Green precipitate
Copper (II), $Cu^{2+}_{(aq)}$	Pale blue precipitate, insoluble in excess	Blue precipitate, soluble in excess, giving dark blue solution	Light blue precipitate turning black on heating
Iron (II), $Fe^{2+}_{(aq)}$	Dirty green precipitate, insoluble in excess	Green precipitate, insoluble in excess	Mud green precipitate, darkens in the air
Iron (III), $Fe^{3+}_{(aq)}$	Red-brown precipitate, insoluble in excess	Red-brown precipitate, insoluble in excess	Red-brown precipitate, CO_2 evolved
Lead (II), $Pb^{2+}_{(aq)}$	White precipitate, soluble in excess	White precipitate, insoluble in excess	White precipitate
Magnesium, $Mg^{2+}_{(aq)}$	White precipitate, insoluble in excess	White precipitate, insoluble in excess	White precipitate
Manganese (II), $Mn^{2+}_{(aq)}$	Off-white precipitate, insoluble in excess, darkens in air	Off-white precipitate, insoluble in excess	white precipitate rapidly goes brown in air
Zinc, $Zn^{2+}_{(aq)}$	White precipitate, soluble in excess	White precipitate, soluble in excess	White precipitate

Test to distinguish between Pb^{2+} ions and Al^{3+} ions: Pb^{2+} forms a yellow precipitate with the iodide ion and the Chromate ion. The Al^{3+} ion does not react with either anion.

Also $PbCl_2$ is soluble when hot and insoluble when cold.

171

Flame tests

Cations have stable electronic configurations. A Bunsen flame in the laboratory therefore cannot provide sufficient energy to bring about further ionisation.

However, when a cation is heated, an electron absorbs energy, becomes excited and moves into higher energy levels where it is unstable. As it returns to its ground state in the ion, it emits the energy absorbed, as visible light of specific wavelengths and colours.

Each ion has characteristic wavelengths absorbed and emitted, and can be identified by the colour of the flame.

Method

Clean platinum or nichrome wire or a glass rod, by alternately placing it in concentrated hydrochloric acid and a clean blue non-luminous Bunsen flame.

Dip the wire in the test solution and place it in the non-luminous Bunsen flame

Record the colour of the flame.

Colour of flame	Ion most likely
Red	Li^+
Crimson	Sr^{2+}
Brick red	Ca^{2+}
Yellow/orange	Na^+
Apple green	Ba^{2+}
Bluish/green	Cu^{2+}
Lilac	K^+ (Na^+ is often an impurity. To filter out the yellow of the Na^+ion, the flame should be viewed through a cobalt-blue glass)

Tests for anions

Ion	Reaction
Carbonate, CO_3^{2-}	CO_2 liberated by dilute acids
Chromate (VI), $CrO_4^{2-}{}_{(aq)}$	Yellow solution turns orange with $H^+{}_{(aq)}$; gives yellow precipitate with $Ba^{2+}{}_{(aq)}$ gives bright yellow precipitate with $Pb^{2+}{}_{(Aq)}$
Chloride, $Cl^-{}_{(aq)}$	Gives white precipitate with $Ag^+{}_{(aq)}$ (soluble in $NH_{3(aq)}$); gives white precipitate with $Pb^{2+}{}_{(aq)}$
Bromide, $Br^-{}_{(aq)}$	Gives cream precipitate with $Ag^+{}_{(aq)}$ (partially soluble in dilute $NH_{3(aq)}$ soluble in conc. $NH_{3(aq)}$); gives yellow precipitate with $Pb^{2+}{}_{(aq)}$
Iodide, $I^-{}_{(aq)}$	Gives yellow precipitate with $Ag^+{}_{(aq)}$ (insoluble in $NH_{3(aq)}$); gives yellow precipitate with $Pb^{2+}{}_{(aq)}$
Nitrate, $NO_3^-{}_{(aq)}$	$NH_{3(g)}$ liberated on heating with $OH^-{}_{(aq)}$ and Al foil; Cu + conc. $H_2SO_4 \rightarrow$ Brown NO_2 + blue solution
Nitrite, $NO_2^-{}_{(aq)}$	With $OH^-{}_{(aq)}$ and Al foil; NO liberated by dilute acids (colourless NO $\rightarrow$ (pale) brown NO_2 in air)
Sulphate, $SO_4^{2-}{}_{(aq)}$	Gives white precipitate with $Ba^{2+}{}_{(aq)}$ or with $Pb^{2+}{}_{(aq)}$ (insoluble in excess dilute strong acid)
Sulphite, $SO_3^{2-}{}_{(aq)}$	SO_2 liberated with dilute acids; gives white precipitate with $Ba^{2+}{}_{(aq)}$, soluble in dilute strong acids

Tests for gases

Gas	Tests and Results
Ammonia, NH_3	Colourless, pungent smell, turns moist litmus paper blue, produces dense white fumes ($NH_4Cl_{(s)}$) with $HCl_{(g)}$ from glass rod dipped in $HCl_{(aq)}$
Hydrogen Chloride, HCl	Colourless, odourless, turns moist blue litmus red, fumes in air, produces dense white fumes of($NH_4Cl_{(s)}$) with $NH_{3\ (g)}$
Carbon dioxide, CO_2	Colourless, odourless, turns moist blue litmus paper pink, produces a white precipitate with $Ca(OH)_{2(aq)}$, limewater. Precipitate dissolves in excess lime water.
Chlorine, Cl_2	Yellow-green, pungent smell, turns moist blue litmus paper red then bleaches it.
Hydrogen, H_2	Colourless, odourless, puts out a lighted splint with a popping sound.
Oxygen, O_2	Colourless, odourless, relights a glowing splint.
Sulphur dioxide, SO_2	Colourless, pungent smell, turns moist blue litmus paper red, turns orange $K_2Cr_2O_7/H^+$ green and purple $KMnO_4/H^+$ colourless.

Appendix II
Periodic Table of the Elements

The Modern Periodic Table of the Elements

Appendix III

Answers to Selected Questions

Exercise 1.2 c.	47.93
Exercise 2.2 a.	$2.25\ 10^{10}$ years
Exercise 7.1 a. ii)	C_4H_8
Exercise 7.1 d.	C_3H_6O and $C_6H_{12}O_2$
Exercise 7.1 a. iii)	120 cm^3 NH$_3$ and 20 cm^3 excess H$_2$
Exercise 7.2 c.	89.6 cm^3
Exercise 9.3 a.	- 282.99 kJ
Exercise 10.1 c.	0.22 dm^3
Exercise 10.2 b.	7.85 dm^3
Exercise 10.2 c.	0.20 mole
Exercise 11.2 c. ii)	8.26
Exercise 11.2 c. iii)	8.25
Exercise 12.1 c.	0.04
Exercise 12.1 d.	mol^{-1}dm^{3}s^{-1}
Exercise 13.2 b i)	1.60 V

Appendix IV

Further Reading References

1. Hill, G. C. and Holman, J.S., (1989), Chemistry in Context, ELBS.

2. Cambridge University, (1998), Cambridge Modular Sciences, Cambridge University Press.

3. Brady, J. E, Holum, J. R, (1988) Fundamentals of Chemistry 3rd Edition, John Wiley and Sons.

4. The Modern Periodic Tables of the Elements. Website: http://www.aplustopper.com/modern-periodic-table-significance/ .

www.ingramcontent.com/pod-product-compliance
Lightning Source LLC
Chambersburg PA
CBHW041704210326
41598CB00007B/525